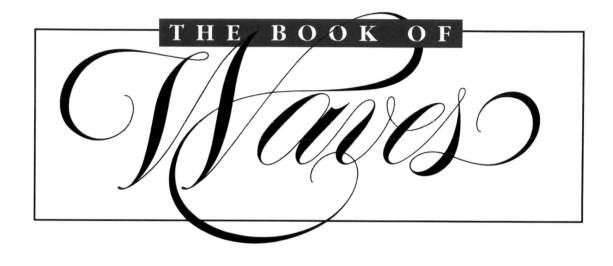

THE BOOK OF

Waves

I

ISBN 1-57098-168-X

Library of Congress Catalog Card Number 97-67690

Copyright © 1997, 1989 Arpel Graphics, Inc. and Surfer Publications
Photography Copyright © 1997, 1989 Surfer Publications
Text Copyright © 1997, 1989 Surfer Publications

Published by ROBERTS RINEHART PUBLISHERS
6309 Monarch Park Place, Niwot, Colorado 80503

Distributed to the trade in the United States and Canada by
Publishers Group West

Published in Ireland by ROBERTS RINEHART PUBLISHERS
Trinity House, Charleston Road, Ranelagh, Dublin 6

Published in England, Scotland and Wales by
ROBERTS RINEHART PUBLISHERS, Airlift Book Company, 8 the Arena
Mollison Avenue, Enfield, Middlesex, England EN3 7NJ

Published in Australia and New Zealand by ROBERTS RINEHART PUBLISHERS,
Peribo PTY Ltd., Mount Kuring - GAI, NSW 2080 Australia

Printed in Belgium

THE BOOK OF Waves

Form and Beauty on the Ocean

With Text by Drew Kampion

Project Directors: Patrick O'Dowd & Steve Pezman

Editorial Director: Drew Kampion

Art Director: Jeff Girard

Photography Editor: Art Brewer

Illustrator: Phil Roberts

ROBERTS RINEHART PUBLISHERS

BOULDER, COLORADO

ACKNOWLEDGMENTS

To Tom Servais, Rob Gilley, Jody Kirk, Jeff Divine, Denise Bashem,
Chris Lyons, Joe Mickey, Susan Kampion, Cathy, Sarah, and Trevor Girard,
Anne O'Dowd, Jerry Samuelson, Willard Bascom, George Downing,
John Severson and the National Aeronautics and Space Administration.
Without their help this book would not have been possible.

Waves

Thou glorious mirror, where the Almighty's form
Glasses itself in tempests; in all time,
Calm or convulsed—in breeze, or gale, or storm,
Icing the pole, or in the torrid clime
Dark-heaving—boundless, endless, and sublime—
The image of Eternity—the throne
Of the Invisible; even from out thy slime
The monsters of the deep are made; each zone
Obeys thee; thou goest forth,
   dread, fathomless, alone.

—Lord Byron

IV

V

VII

VIII

The sea! the sea! the open sea!
The blue, the fresh, the ever free!

—Bryan Waller Proctor

XI

XII

The tossing waves, the foam,
  the ships in the distance,
The wild unrest, the snowy, curling caps—
  that inbound urge and urge of waves,
Seeking the shores forever.

—Whitman

A Meditation On Breaking Waves

*Out of the cradle
endlessly rocking…*

*The white arms
out in the breakers
tirelessly tossing,*

*I, with bare feet,
a child,
the wind
wafting my hair,*

*Listen'd
long and long.*

—Walt Whitman

Waves: pulses of energy, echoes of power, children of the struggle between ocean and atmosphere, endless song of circulation. Vibrations put into motion by the harmonic resonance, the insistent music of the spheres, ocean waves are the ongoing signatures of infinity, eternity and the miraculous. They are the gifts of the invisible to the visible (energy into air into liquid), the voices of intelligent silence, the incarnation of the molecular soul in corporeal reality. As God breathed life into the raw form of Adam, so the wind breathes life into the inert ocean, raises it into a running rhythm, lifts it up out of itself and, finally, transforms the sea into the spectacular living glory of breaking waves.

There, at the meeting of land and sea, ocean waves reveal their essential meaning; deliver their messages. As spectacularly varied as snowflakes, more powerful than avalanches, as relentless as the pull of the moon or the grip of the sun, the ocean—as expressed in its breaking waves—is a limitless source of adventure, mystery and sensual richness.

We sit on the shore and watch the waves moving incessantly toward us (forever and ever throughout all lifetimes) and we marvel at their beauty and thrill to their surprises, while we're soothed (forever and ever) by their cadence. We sit marvelling on the shore, and the waves advance to meet us. They materialize out of the general abstraction of the ocean, climb into increasing definition, then feather, plunge and (ever more slowly) rush or push or crawl ashore.

In the absolute beginning, as infants in the womb, everything came to us as waves, advancing through the primal liquid to acquaint our senses with sublime echoes of an invisible world beyond the realm of our comprehension, beyond the smaller space we knew, making that familiar life of the womb by this mysterious contact somehow more infinite, somehow more momentous, poised on the brink of a greatness, an unknown significance.

So that even now, who could not sit and watch the sea and the breaking waves for hours…for days…forever? Never ending, untiring the march shoreward.

Rank upon rank of waves, leaping, curling, plunging. Each upon each, and absolutely infinite the variety. No two in all the millenia curling over exactly alike. Each one individual. Each one yet another unique expression of the complex interweavings of natural law. Each one an identity. Each part of one an identity. The inexhaustible creativity of it almost overwhelming.

Rank upon rank, echoing that invisible, all-powerful source that all form calls out from. Relentless and utterly reliable, terrible and beautiful.

No doubt men of all times have stood upon the shore and lost themselves in the thunder of the shorebreak, the cadence of the sea, harkening completely to that synchronous metre deep within the human psyche.

Or maybe not so deep. The noted oceanographer, Blair Kinsman wrote: "Phenomena do one of three things: they stand still, in which case the problem is to evaluate a constant; they grow, in which case we try to find a growth law; they oscillate, in which case we have a wave problem." So Kinsman posited a short but potent list of fundamental physical actions—mega-verbs, if you will—that give some indication of the primary importance of wave phenomena in our realities. Indeed, we are all at sea in this life, borne on an infinite network of complex oscillations. No part of our life is free of waves. We are all surfers, are we not?

Aren't we all drawn to the ocean waves? We the watchers, the voyeurs, the admirers, the seekers of that specially charged atmosphere that bathes the holy beach where ocean wave meets solid ground and gives up its accumulated life force in a powerful expression of consummation…one after another, eternally (or so it seems to us), electrifying that dynamic surf zone with charge after charge of blown energy set free to hover there.

Is it any wonder humanity flocks to the shore? We explain it in visible, tangible, sensational, reasonable ways—the cool, the water, the fun, the sun—when really it's the irresistible magnetic attraction to the energy of the place.

Yet the waves, those transient objects of our fascination, both invite and threaten. These wonders of nature are as well known for their destructive force as they are for their beauty. Ask the thousands or millions who've been consumed by the oceans and their waves. Ask them about its beauty. Ask them about its many moods.

Who would venture to guess how many millenia it took for a man to finally dare approach the ocean waves with his first rude craft? Who knows how many millenia more it took for man to actually challenge the surf, to leap on the back of the wild steed and ride until it collapsed, exhausted, up onto the shore? A long time, I'll wager. Millenia aplenty.

Unless, for instance, man's relationship with the sea and the waves was given to him—breathed into his essential consciousness by some force, some power, some other mind. Think about it: How likely is it that any conceivable evolution would have taken man down a critical path that would lead him eventually (and in our time) into the hollow pocket of a 30-foot wave…to ride in the belly of the beast…to slide through the maelstrom of a collapsing cathedral of water with the composure of a matador, with the exhilaration of a wild-haired youth, with the primitive awe of an elemental man, with the futuristic aplomb of a science-fiction comic book hero come to life, incarnate in his most bizarre predicament of all?

Picture a twister lying on its side in the ocean, pulling the water over and around itself like a ragged, raging blanket, its wider open end revealing a swirling hollowed-out cavern of spitting, hissing power, rolling like a giant tunnel toward the shore. Imagine the kind of man who'd want to put himself in the eye of this thundering killer, who—like a matador to an avalanche—would wish to stand cooly in the raging glass fist, while time grinds nearly to a halt and where the secrets of immortality are whispered to those who have ears to hear.

Perhaps it is as the German poet Rainier Maria Rilke has said, that we are "the bees of the invisible," here to harvest with our senses the realities of the physical world around us. Why? To function as the eyes and ears (and all the senses) of a being or beings that cannot directly experience this world or this reality without us. Maybe receiving and transmitting the essential knowledge of this world—more divine than we can suspect—is the central reason for our existence.

So said Rilke, and so can we ourselves ponder the imponderable questions: Why our attraction to the sea and the waves? What is the lure that seems so basic, so essential? And could it be that our play (and our crazed risk-taking) in the waves has some "larger" purpose? Is there something miraculous afoot here that we can only vaguely sense?

In fact, the surface of the sea (and its shore, and even its depths) is at the interface of several different worlds. Not different worlds philosophically, but different worlds *in reality*. Interestingly, the form of communication between these worlds is waves. The medium changes, the form and the message remain the same.

Start with the sun, worshiped through antiquity as the Great Cause of all life, now worshiped or reviled only for its observable physical effects. The sun is taken for granted as surely as our next breath.

Yet the essential relationships perceived by the

ancients between our world and the sun are still as valid, still as active, still (to this day) as real. Our blindness to these relationships does not diminish them. How could it?

The sun communicates its energy—light and heat, yes, but some other basic energy of life too—through "rays," waves of influence, waves of energy, waves of communication. The rays from the sun excite and energize the atmosphere of the earth, awakening it to movement, to flow, to rhythm, to life. The wind becomes the voice of the atmosphere, but its words have come from the sun. The wind speaks the message of the sun to the sea, and the sea transmits it on through waves. The wave is the messenger, water the medium, and the message (like all messages) is energy—the energy of the sun…the energy of life.

Wind whispers the message—or shouts it or roars it—and the echo flies out through the water, expanding, aroused, relentless. At first, ripples (like voices) vibrate into life, then they overtake and merge with one another, gathering into larger waves, larger voices. More and more they overtake, join and merge together until countless ripples (like voices) have organized into massive bands of communication that radiate out and sweep across and through the vast body of the world's oceans.

Waves, transient objects of our fascination, both invite and threaten. Who could not sit and watch for hours? Children watch from atop a World War II bunker on Christmas Island. Photo by Bernie Baker.

*Who hath
desired the Sea?*

*—the sight
of salt water
unbounded—*

*The heave
and the halt
and the howl
and the crash
of the comber
wind-hounded?*

—Rudyard Kipling

It is no accident, then, that the shore has become our favorite meditative perch. No accident that the incessant rhythms of the sea calm us, bring us back into phase, reacquaint us with ourselves. No accident that we compete with each other for an ocean view. But it's not really some "view" we're after. It's that ancient, exquisite, powerful message of the breaking waves.

I have stood on the beach on Oahu's North Shore with the ground under my bare feet shaking with the impact of giant surf on the "cloudbreak" reefs a mile or so out to sea. I stood there and watched the giants rise up out of the rugged horizon, arch up into magnificent walls (I had no way to estimate their size), then curl over in slow-motion silence to the count of one…two…three…four—so that now I had some idea of how big they were—leaving behind a somehow frightening sweep of blue-white mist that drifted back to the water like a diaphanous veil out from under which some beauty has just inexplicably vanished.

I stood there as the raging wall of blasted white water advanced on the shore with no sign of weakness or slowing, a wall of white water so huge that the new giants farther to sea could barely be seen fringing above. And then the white water, driven ever forward by too much force, by force uncontainable, was absorbed back into the ocean itself, as the giant wave steadily and ominously began to reform closer to shore, growing larger and larger as it began to loom closer to the reef and the sand banks just off shore.

I stood there at the high edge of where the last giant had lately smoothed the golden sands, looking down a long, steady slope to the churning mass of the sea. I knew there was nothing to stop the great momentum that was to come, yet I could not move. I didn't want to. The words in my head urged retreat, but the great wave—now arched like a cobra over the inside reef, beginning to curl over in a twisted, muscular mass of storm-brown water—absolutely compelled my attention. So I stood and watched it torque and grind and explode into another seething wall of angry white water (though its whiteness was not of any particular purity; it was a whiteness by contrast) and then it was closing on the shore, first draining, then swelling the ocean's rim, until finally it drove up the beach toward me in the form of a powerful, sweeping surge. I had never seen white water come up the beach so fast.

It was a broad beach—maybe a hundred yards—but the wave covered it in three or four seconds…and still I could not move. I stood there with my legs apart, and the water surged up the beach and around me and past me; the sensation of the racing water around my legs was powerful and exhilarating.

The water rocked me, but it didn't topple me. It was then that I turned and saw how it had run up over the ledge to the house behind, running under it, encircling it. The trick, in the wave's retreat, was to keep out of the way of the things that it was taking back to sea with it—coconuts, a lawn chair, a volley-ball, a surfboard (which I managed to retrieve). And then the wave was gone, the area around me hissing with dying foam, and the sweep of golden sand down to the water's edge was washed clean.

Other times, I've stood on that same shore, looking through binoculars a half-mile down the beach into the

curling tube of the famous Banzai Pipeline wave, watching from a clinical distance the wrenching suddenness with which the massive ground swells heaved high over the coral reef…watching the racing wall of water go completely vertical, then throw out a six-foot-thick lip from the top that pulled the wave into a cylinder big enough to surround a semi-truck. I've watched surfers slide deftly over the teetering cornices of waves like that, slide right down the face on the ragged edge of control, then take aim and come flying down the barrel toward me. Most are shot out alive.

I've stood on the beach in Western Australia and watched the waves there, remarking to myself that the different water of the different oceans makes the waves somehow different. Different in more than just water color or power—more a difference in quality. And the surfers there guard their waves jealously and ask you where you're from and if they can please have the roll of exposed film from your camera.

I've watched monster standing waves off Portland Bill in the English Channel, where strong currents and strong winds in direct opposition create a wild, nightmarish seascape that you have to watch from shore to truly appreciate…and be able to tell about later. These are the kinds of seas that spawn true lore. At times the local folk have stood on the low cliffs above the raging sea like spectators in a gallery, watching boats going down with all hands.

Off Lighthouse Point in Santa Cruz, California, I've seen a tremendous stack of smoking hump-backed peaks looming up out at the second and third reefs of Steamer Lane, back-lit to an almost incandescent green by the late afternoon sun. I've seen seals wriggle into the fall lines of these beauties and catch a free ride into the cove with all the apparent glee of any other surfer on the planet.

I've watched big, savage tubes uncoiling right along the rim of a reef planted with razor-sharp red fire coral in Guam, with surfers slotted back in the barrel, walking a fine line with disaster. Even so, they can't resist the lure. They do it every day.

And many an afternoon I played in the waves at Makaha on the leeward shore of Oahu, where perfect four- or five-foot waves would peel off along the reef toward the big bowl of steep beach. And then the reflection of a wave that went before would rush out to meet the one approaching and, like sumo wrestlers colliding belly to belly with a wild slap, a fan of transparent water would be flung up across the setting sun.

I've spent many a late afternoon afloat on a surfboard, lying or sitting on it out where the waves form, waiting for a good "set" to come through with a perfect wave to end the perfect day.

It's a great feeling out there with the waves, especially on those days when the swell is clean and the sea is glassy and the fiery colors of the dying sunset burnish the world around you with golds and oranges and purples.

And then a dark line lifts up out of the ocean seaward and a wave moves silently toward you—a wave that has come five hundred or five thousand miles—and you turn your board toward shore and drive your hands alternately into the cool water, gathering speed till the wave starts to lift you…lift you…and then you're sliding down the smooth face, turning ahead of the curling peak to speed across the hollowing wall. The feelings and sensations created are euphorically beyond words.

This is how I explain the ocean and its waves to my two-year-old daughter: This is the ocean water. This is where the fish lives, where the shark hunts and the whale plays. Those are waves of water—one and another and another and another and another. More and more. On and on. Like a clock. Like time. Beautiful ocean water…beautiful waves…beautiful time.

And this is how I explain the ocean and its waves to my five-year-old son: This is the ocean. It's the biggest part of nature that we can get to that's still alive. We have to try very hard to keep it alive. And these ocean waves are perhaps the most incredible things on the planet. I used to lie on the beach and watch them for hours and hours at a time. I have never gotten tired of watching them. It's almost as if they're trying to tell me something. Or that they *are* telling me something, and I'm trying hard to *hear* it. You've probably noticed how much I like to surf or bodysurf in the waves; it makes me realize that I'm alive in a very strange and wonderful world because nothing in this world is more wonderful or more strange than waves. I hope we'll be able to play in them and ride them together when you learn how to swim, which I hope will be very soon.

And this is how I explain them to myself: Everything is waves. The universe of space and matter is charged with energy, and this energy is organized by God or by forces far greater than ouselves into the pulsations we call waves. Waves of energy. Like echoes of the heartbeat of the absolute being, waves give expression to the divine will. They give form to the universe.

The passage of energy through matter organizes matter, and waves pass through everything—steel, stone, flesh and blood and water and air and space alike. Waves are the imprint, the signature, not only of life, but of existence itself.

Waves penetrate, pass through and shape everything, but the medium is not the message. Space, air, water, blood, flesh, stone and steel are not the messages. The messages are what is contained in each wave, and the message is energy.

Light waves emanate out of the sun and the stars and their reflections (like satellites or fire or electrical spark or glowing minerals or luminous fish and bugs). Seismic waves move through solids, liquids or gases. Out on the surface of the ocean, it is the movement of the atmosphere, the wind, rubbing against the water that feathers the surface, that coaxes the ripples into their gentle side-by-side expansion, then forces wavelet upon wavelet till they gather, amplify, swell and expand into giants that seem to groan with potential power as they sweep out across the great plain of the ocean, running free as if nothing could ever stop them. Until, surprisingly, they trip over some buried coral reef, lurch forward and take the fatal fall—that fall that thrills our eyes and electrifies our senses—upon that net of beaches which, from the beginning, was fated to trap the potential of every wave that's created.

Once upon a time, in Utah of all places, the whole tale was revealed to me in a 200-yard-long, shallow strip of water alongside the Interstate. There was a solid 40-knot wind blowing straight out of the north, right down the length of this small pond, if you could call it a pond, since it was actually nothing more than the traces of a Caterpillar D-8 or some other heavy machine that had lowered its blade to excavate this shallow, purposeless cut which had since become pleasantly bordered in meadow grasses and filled to the brim with rainwater.

On this particular day the sky was magnificent, blown clear and dry by the wind, and the water in this strip of pond was as deep and rich and thick a blue as a wet smear of fresh aquamarine oil paint. But the wind told a grand tale on that minute parcel of water.

At the north edge of the pond, the winds swept the grass bendingly out over the glassy, green-blue, mirror-smooth surface. The reflection of the grass, the sky, the occasional cloud there was near to perfect. You could see, then, a foot or two out where the atmosphere—the wind—was having its first effect. There was the slightest suggestion of distortion at first, a nuance of texture, almost a mirage you might have thought, except immediately after there came a deepening suggestion, an affirmation of the movement, a tendency to a pattern that somehow suggested an approaching chaos. Yes, energy threatening established order, swinging reality toward some new and unseen balance…nature resolving its differences in the only lawful way possible.

Then the pattern deepened as thrilling texture became eloquent rippling, and the yang front sides of the advancing microwaves went almost jet blue-black, chased (each one) by its yin half toward oblivion. For there was an oblivion swiftly approaching as latter ripple overtook former ripple (how it happened, I don't know) and in the overtaking both were enlarged. And then that ripple overtook another slower ripple…redoubling, redoubling…until at some dizzying height around three inches their crests began to rip apart, bursting into blossoms of white, chasing across this diminutive pond like creatures with their hair blowing, leaping forward—still pushed from behind by that steady, strong 40 knots, yet somehow chasing ahead now on their own, caught up in a stampede, a momentum of some passion, some joy (I sensed), some passion to arrive. But where?

Ah, yes. The waves—because they were really waves by now, five or six inches high, clearly delineated, formed, individual and relatively powerful—sensed the approach of the opposite shore…the bottom gradually shoaling (where the D-8 had sculpted its first slice into the soil), forcing the energy up, pushing the wave up out of itself, creating feathering, fringing lines of surf that rose, hollowed, folded and spilled up onto the distant beach a lifetime away from the other side. One after another, each after each. All different, and all the same. The waves were born, they lived and they died. Lives on the pond.

And still the wind rubbed its cool body across the water—lured, stroked, gathered and chased the surface into waves—precisely akin to the process far out to sea, far beyond the reach of our straining eyes, where the waves that finally dissolve away at our feet were born. But how?

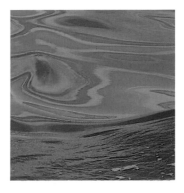

Reflections in a boat-wake wave in the channel between the Hawaiian Islands of Lanai and Maui. Photos by John Severson.

Origins

She will start from her slumber
When gusts shake the door;
She will hear the wind howling,
Will hear the waves roar.
We shall see, while above us
The waves roar and whirl,
A ceiling of amber,
A pavement of pearl.
Singing: "Here came a mortal,
But faithless was she!
And alone swell forever
The kings of the sea."

—Matthew Arnold

THE LIFECYCLE OF OCEAN WAVES

Under heaven nothing is more soft and yielding than water.

Yet for attacking the solid and strong, nothing is better;

It has no equal.

—Lao Tsu

Figure 1—An ideal wave: *This familiar sinusoidal pattern is echoed throughout nature, although this simplified model exists only in theory or in the laboratory.*

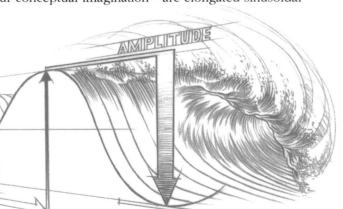

We are surrounded and influenced everywhere by waves. From the radiations of light and color, to the sounds that vibrate through our atmosphere, to the cycles of the tides, and of night and day, and of the movements of our lives—it seems that everything comes in waves, or as cycles moving within waves.

Clearly, wave action is the fundamental way in which energy is transported and transmitted in this world. Waves are an expression of the universal rhythm that orchestrates and propels all creation and the development of life on earth. Perhaps this is why the contemplation and study of ocean waves is so attractive, so compelling.

Ocean waves are among the earth's most complicated natural phenomena, yet when we picture waves in the abstract, our minds might conjure an image of the perfect concentric ripples that echo the point of entry of a pebble into smooth pond waters. Those waves—the ideal waves of our conceptual imagination—are elongated sinusoidal oscillations *[Fig. 1]*, and although they do exist in relatively pure form in controlled conditions, they are not likely to be found in the more complex ocean environment *[Fig 2]*. This is why waves are usually studied in laboratory tanks, where a single train of waves can be generated and where the mechanics of wave motion can be isolated and simplified.

Ocean waves and laboratory waves share the same basic features: a crest (the highest point of the wave), a trough (the lowest point), a height (the vertical distance from the trough to the crest), a wave length (the horizontal distance between two wave crests, and a period (the time it takes for a wave crest to travel one wave length) *[Fig 3]*.

Standing on a pier or jetty, or sitting astride a surfboard, the swift approach of an ocean wave gives the impression of a wall of water moving in your direction. In actuality, although the wave is moving toward you, the water is not. If the water were moving with the wave, the ocean and everything on it would be racing into the shore with catastrophic results. Instead, the wave moves through the water, leaving the water about where it was.

Spread a blanket on the floor. Kneel at one end and take the edge of the blanket in your hands, then slowly snap waves down its length. The blanket doesn't move, the waves ripple through it. The energy crosses the blanket in an oscillating wave pattern, diminishing (or decaying) as it moves toward the opposite end.

An ocean wave passing through deep water causes a particle on the surface to move in a roughly circular orbit, drawing the particle first toward the advancing wave, then up into the wave, then forward with it, then—as the wave leaves the particle behind—back to its starting point *[Fig. 4]*.

Because the speed is greater at the top of the orbit than at the bottom, the particle is not returned exactly to its original position after the passing of a wave, but has moved slightly in the direction of the wave motion.

The radius of this circular orbit decreases with depth. In shallower water the orbits become increasingly elliptical until, in very shallow water—at a beach—the vertical motion disappears almost completely.

Its final destruction in shallow water culminates the three phases in the life of a wave. From birth to maturity to death, a wave is subject to the same laws as any other "living" thing, and—like other living things—each wave assumes for a time a miraculous individuality that, in the end, is reabsorbed into the great ocean of life.

The Origins of Waves

Undulating ocean surface waves are primarily generated by three natural causes: wind, seismic disturbances and the gravitational pull of the moon and the sun. Oceanographers call all three "gravity" waves, since once they have been generated gravity is the force that drives them in an attempt to restore the ocean surface to a flat plain.

There are other waves, too, in the ocean. At the boundaries of cold and warm currents, submarine streams of different density undulate past each other in slow-moving "internal" waves. The evidence of internal waves can sometimes be seen in calm conditions since their currents affect the reflectivity of the ocean's surface, producing alternating areas of glassy slickness and ruffled texture.

Although significant seismic-wave disturbances (tsunamis) are still popularly known as "tidal waves," the term more accurately describes the daily cycles of high and low tides. The greatest ocean waves of all—with a period of 12 hours and 25 minutes and a wave length of half the circumference of the earth—these colossal oceanic bulges travel around the world at up to 700 or 800 miles per hour. The tides are created when the massive gravitational pulls of the moon and the sun actually lift the oceans while the earth rotates by underneath. The crests of these waves are the high tides, the troughs low tides.

One unusual tidal wave phenomenon is a "bore," the sudden surge with which the incoming tide arrives in some parts of the world. Bores occur in streams or rivers (like Britain's Severn River) or bays (like the Bay of Fundy in Nova Scotia) with funnel-shaped shores and shoaling bottoms where tidal ranges are high. If the incoming tide is retarded by friction in the shallowing water until it moves more slowly than the outgoing current, the tidal surge can build up into a turbulent crest. The resulting bore wave may drive up a narrowing passage with great energy and force.

Augmented by a west wind and spring tides, the bore on France's river Seine (called the mascaret) has been known to arrive at Paris as a great wall of water moving at high speed. One report claims a 24-foot-high wall of water traveling 15 miles per hour. This is the tidal bore that drowned Victor Hugo's newly married daughter and her husband, who were caught while sailing on the river in front of Hugo's home.

The other "tidal waves"—seismic sea waves, or tsunamis—are "impulsively generated" waves, most commonly by earthquakes, volcanic eruptions or massive underwater landslides. The waves created by such abrupt forces can be very long and low with periods between crests of up to ten minutes and wave lengths as long as 150 miles. Yet the waves

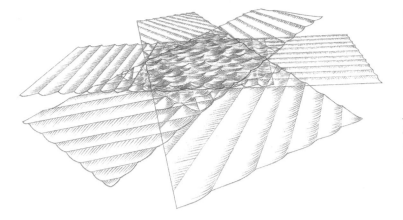

Figure 2—The surface of the sea: The interaction of many simple sine wave patterns creates a sea.

are usually only a foot or two high in deep ocean water, and the slope of a tsunami wave face can be so gradual that ships at sea are unlikely to even notice its passage.

Tsunami waves travel extremely fast—about 500 miles per hour in the mid-Pacific—and the energy they transmit can be massive indeed. But as stealthy and swift as they are through the ocean, these seismic waves assume a completely different character when they encounter a shoaling bottom.

The most notable example of the destructive power of an explosively-generated tsunami is the volcanic eruption in 1883 of the northern portion of Krakatoa, an island located in the Sunda Strait between Java and Sumatra. Some five cubic miles of lava, pumice and ash were blown out in a massive and sudden eruption, leaving a 900-foot-deep crater where a 700-foot-high land mass had been. The blast was heard in Madagascar 3,000 miles away. Although immense physical destruction was caused by the explosion, the real catastrophe was caused by the resulting tsunami, which ranged from 60 to 120 feet high. Some 300 towns and villages on the shores of nearby islands were destroyed; over 36,000 people were killed. The gunboat *Berouw*, anchored off Sumatra, was carried nearly two miles inland, and gauges in France and Britain recorded a rise in the sea level.

In 1960, a violent earthquake in Chile (magnitude 8.5) caused a great subsidence of the undersea fault that parallels the coast there, generating a catastrophic tsunami that affected nearly all of the Pacific basin. Australia, New Zealand,

Figure 3—The anatomy of an ocean wave: Whatever the medium they move through, all waves share the same basic physical characteristics.

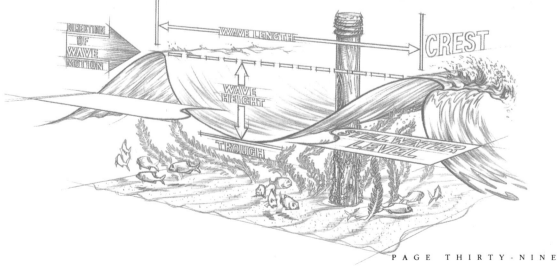

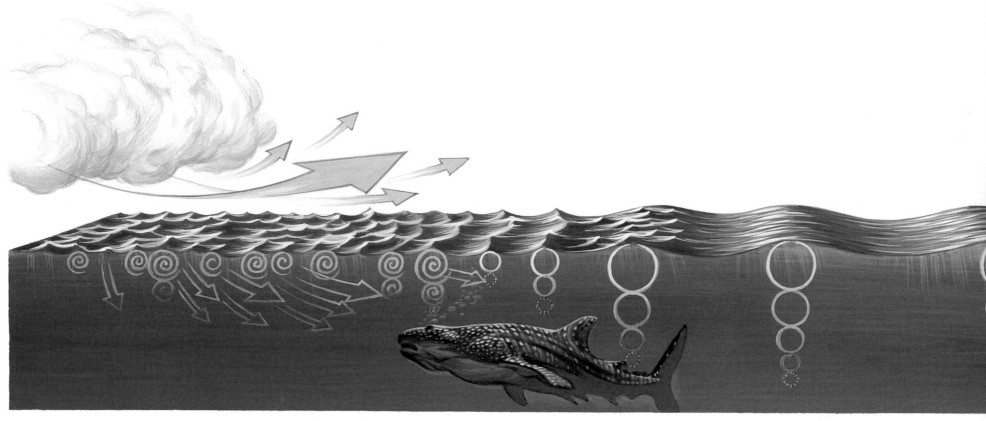

The Lifecycle of Ocean Waves

Genesis: *Winds blowing across the water's surface raise ripples, then chop. If wind strength, duration and fetch are sufficient, a "sea" develops.*

Fetch: *The area over which the wind blows to raise up waves; most (but not all) of the atmospheric energy transferred to the water by frictional forces is concentrated at or near the surface.*

Maturity: *Once the seas leave the fetch area, the locally confused patterns organize themselves into lines of swell that radiate downwind from the area of genesis.*

the Philippines, Okinawa and California experienced significant coastal flooding or damage. Fifteen-foot waves were hurled against Japan, some 9,000 miles from Chile, and the city of Hilo on the island of Hawaii (which had been devastated by a tremendous tsunami as recently as April 1, 1946) was virtually washed away by a series of massive seismic sea waves that began to hit less than three hours after the quake. Hilo has since been rebuilt on higher ground, dedicating the former site—now called "Tsunami Park"—for recreational use.

Although tsunamis are certainly spectacular if you're in the right place at the wrong time, they are relatively rare. And the tides (although they're always with us) are relatively slow to shift and difficult to observe as waves. On a day-to-day basis, wind-generated waves are the most visible to us. Ripples, chop, rough seas or plunging breakers, these are

Figure 4—How particles are moved by waves: An individual molecule of water is displaced in a circular pattern by the passage of a wave, ending up about where it started.

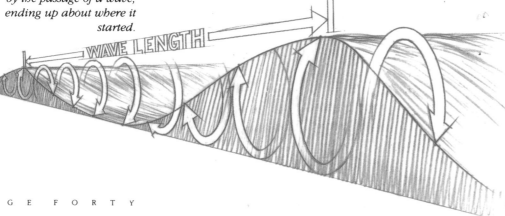

WAVE LENGTH

what we think of when we hear the word "waves," and their source is the movement of air across water.

Wind is the result of solar energy acting on the earth's atmosphere. The great patterns of circulation—the global winds—give rise to the various dynamics of high and low pressure, of calm and storm. Huge North Pacific or North Atlantic or Antarctic systems generate enormous waves. More localized thermal differentials excite the ocean's surface with racing patterns of energy. Smooth coastal waters oscillate gently with the decaying echoes of storms half a world away.

How does the wind make waves? The primary mechanism of wave genesis is the friction between the atmosphere and the surface of the ocean. A puff of less than two knots will raise miniscule wrinkles (called capillary waves) on the surface almost immediately. As the puff dies, these waves quickly disappear due to the resistance of the water's surface tension, which tends to restore the smooth surface. However, when a breeze of two knots or more develops and is sustained for a time, "gravity waves" begin to form as the wind drags across the water. Ripples at first, these waves continue to grow as the wind continues to blow. In fact, it becomes increasingly easy for the wind to transfer its energy to the water since it can now push directly against the backs of the ripples. The more jagged and uneven the surface, the more there is for the wind to push against. Ripples develop into chop (periods of one to four seconds) until, when the wave length of the chop in a given area stretches beyond five seconds or so, it is called "sea" *[Fig. 5].*

As the waves continue to grow, the surface resisting the wind becomes steeper and higher, making the wind's work

Particle movement: *Waves passing through water cause particles near the surface to rotate in circular orbits. The diameters of these orbits diminish as depth increases.*

Landfall: *As swells begin to be affected by a shoaling bottom, their character begins to change; they begin to slow, the wave length shortens and, when the bottom is shallow enough, they break.*

Breaking waves: *When a shoaling bottom causes waves to become critically steep, they peak up and break; the shallow water no longer allows the complete internal rotation of the water particles.*

Final moments: *The momentum of the plunging breakers pushes water toward the shore, expending the last of the wave energy.*

of transferring energy to the water still more efficient. But there is a limit to how large these waves can grow. Steepness is a ratio of the height of a wave to its length which, it turns out, can't exceed approximately 1:7. This means that a seven-foot-long wave can't have a crest taller than a foot. In fact, the maximum stable profile angle of a wave crest is about 120 degrees. Beyond this point the wave will begin to break into whitecaps.

How large wind waves become is a function of three factors: the strength of the wind (force), the length of time it blows (duration), and the amount of open water over which it blows (the fetch). If the wind is strong enough and blows long enough, waves of considerable size can develop. However, there is a limit to the amount of energy that can be transferred from the atmosphere to the ocean for a given wind strength, and when that limit has been reached, the seas are said to be fully developed or fully aroused. For instance, an accepted mathematical model suggests that if the wind blows at a velocity of 30 knots over a fetch of some 280 nautical miles for at least 23 hours, a fully-arisen sea will be the result, with average waves of 13 feet and the highest waves approaching 30 feet.

Waves generated by the kinds of storms that actually happen seldom need fetches of more than 600 to 700 nautical miles to reach full height. According to oceanographer Blair Kinsman, 900 nautical miles is probably room enough to develop the largest storm waves that have ever been reliably estimated. Occasional open-ocean waves of 40 to 50 feet do occur, he says, but they are not common, and even in the worst storms the run is much smaller.

Kinsman developed an estimate for the "whole ocean" based on a frequency study for wave heights (over 40 thousand extracts) developed by Bigelow and Edmondson in 1947, which seems reasonable:

Wave height	0-3'	3'-4'	4'-7'	7'-12'	12'-20'	over 20
Frequency of occurrence	20%	25%	20%	15%	10%	10%

This would indicate that 45 percent of all ocean waves are less than 4 feet high, and 80 percent are less than 12 feet high. Just 10 percent are over 20 feet. The largest wave ever reliably reported had an estimated height of 112 feet. It was encountered on 7 February 1933, during a long stretch of stormy weather, by the *U.S.S. Ramapo* in the North Pacific.

In all their immense variety, waves give texture, motion and character to the world's seas. Having been aroused by the wind and gathered into radiating bands of energy, waves can travel great distances, carrying nearly intact their messages from the sun.

Maturity

Once a pattern of waves radiates free of the winds that created it, the confused chaos of apparently random sea organizes itself into even lines of "swell." The original wind waves decay, and their energy is consolidated into waves of greater length and increasing speed.

As waves increase in height, wave length also increases. In fact, even after wave height has stabilized, the lengths may continue to increase. As a rule, a ten-second period is the dividing line between sea and swell, although there is naturally some overlap. Sea is shorter in wave length, steeper, more jagged and more confused than swell. Like those ripples in the puddle, the crests of open-ocean swell are more rounded and regular, having absorbed the energies of many decaying wave oscillations into relatively unified and orderly packages capable of traveling great distances.

Swell moves across the open ocean in trains of waves of similar period that radiate downwind from a wind source. Responding to the downward force of gravity, the lines of swell spread their forms, lose some height and distribute some of their energy sideways, lengthening the wave front as they expand away from their source. The process is called maturing.

Most open-ocean waves are deep-water waves. This means that the depth of the water the waves are traveling in is greater than half the distance between crests (the wave length). Waves moving in water shallower than half their wave length are known as shallow-water waves. The wave lengths of some seismic waves (generated by earthquakes, for instance) are so long that for them even the deepest ocean is shallow. The dynamics of shallow-water waves are affected by the ocean's bottom, whereas deep-water waves can be studied independent of this influence.

In deep water, the wave length (L) in feet can theoretically be related to the period (P) in seconds by the formula: $L = 5.12\,P^2$. Actual wave length has been found to be somewhat less than this for swell and about two-thirds the value for sea. When waves leave the generating area and continue to move on as free waves, the wave length and period continue to increase while the height decreases. Speed also increases as period increases and is virtually independent of wave height and steepness [Fig 6].

This theoretical description of the relationship between wave speed, wave length and wave period describes deep-water waves only. The relationhip between these characteristics in shallow or shoaling water can be quite different.

Storm waves in the North Atlantic average about 500 feet long; in the North Pacific they may be a bit longer. In the Antarctic waves spawned in the Roaring Forties can have wave lengths greater than 1,000 feet. Lines of swell can have much greater wave length than waves in sea. Kinsman reports swell with lengths of 1,320 feet in the Bay of Biscay, 2,549 feet on the south coast of England, and, the longest on record, 2,719 feet in the equatorial Atlantic. Where fetches are more restricted, wave lengths are naturally smaller. The longest wave length recorded in the Mediterranean Sea, for example, was 328 feet.

Under favorable conditions, swells move indefinitely in the direction of the originating wind. However, if a swell encounters new winds, the shape and heading of the waves may be altered. A strong enough opposing wind can dissipate the waves entirely, while wind or swell moving in the same direction can have an augmenting effect.

Beach Break Wave
Hossegor, France

Wind blowing from shore can smooth and shape beach break waves into perfect plunging cylinders.

Surging water pushed toward the beach escapes seaward through deeper-water channels.

The presence of a continental shelf some miles out to sea slows many beach break waves and diminishes their energy and power.

Large storms and variations in swell direction produced by seasonal patterns can change the character and configuration of a beach break wave spot in a matter or days or even hours.

OVERVIEW

Water surging toward shore in broken waves sets up a littoral current and deeper-water channel inshore of the primary surf zone.

Shifting banks of sand or gravel create an alternating pattern of breaking waves and open channels.

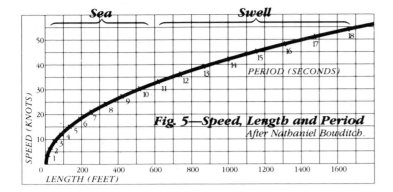

Fig. 5—Speed, Length and Period
After Nathaniel Bowditch.

Because the length, period and speed of waves all increase as the swell moves away from the generating area, it is possible to have a fairly good idea how far away from a point of observation waves were spawned. However, when making the necessary calculations, it is important to know that the time needed for a wave system to travel a given distance is double what it would take an individual wave to go as far. This is because the front wave of an advancing swell gradually disappears, transferring its energy to the following waves. The process is followed by each leading wave in succession at such a rate that the wave train advances at a speed which is just half that of individual waves. The speed at which the wave system advances is called the group velocity *[Fig. 7]*.

Still, for all their apparent symmetry, both theoretical and actual, wind waves are irregular phenomena. Even in trains of open-ocean swell, successive waves can and do differ markedly in height. For instance, in the mathematical model mentioned above, the average wave height created by winds blowing 30 knots for 23 hours over a fetch of 280 nautical miles will be 13.5 feet, however this same model tells us that the "significant wave height" generated will be 21.6 feet. "Significant wave height" is defined as the mean height of the highest one-third of the well-defined waves observed at a given point on the ocean's surface. Usually, as in this example, the significant wave height is about one-and-a-half times the average wave height. However, the average height of the top ten percent of these significant waves will be 27.6 feet. This means that, within a particular wave train, about one wave in a thousand (perhaps one every four hours) will be twice the average size!

One explanation for observed differences in wave height is the interference of one wave train with another. When the peak of one wave synchronizes with the trough of another wave, there is a distinct dampening effect. Conversely, the synchronicity of crests causes wave energies to combine so that the resulting wave can be much larger than either of the two waves that coincided. The swell pattern resulting from the confluence of two or more open-ocean wave trains results in a cycle of larger and smaller waves. Closer to shore this pattern is termed "surf beat" and the larger groups of waves thus created are called "sets."

The most spectacular example of the synchronicity of wave crests (often in combination with other factors) is the phenomenon of "rogue" waves. Rogue waves are statistical probabilities that on rare occasions emerge out of the land of the theoretical to haunt some of the most trafficked sea lanes of the world.

Rogue waves are solitary giants formed out of the convergence of extreme natural forces; they rise to unusual height and mass. Inevitably ships at sea come into contact with some of these wind-generated, gravity-propelled monsters. There are a number of remarkable stories of such encounters in nautical records, but how many other encounters left no tongue alive to tell is food for speculation.

In his authoritative book, *Waves and Beaches*, Willard Bascom cites a number of meetings with rogue waves. A sampler:

"In February 1883, the 320-foot steamship *Glamorgan* out of Liverpool was beating through heavy Atlantic seas at night when it was 'totally submerged by one tremendous wave.' The wave swept away the foremast, all the deckhouses and the bridge (with the captain and seven crew in it). It stove in all the hatches and the engine room was flooded. The ship sank the next morning, and the 44 who escaped in lifeboats told the tale of the one great wave.

"On another Atlantic crossing, the 1,000' *Queen Mary* was serving as a World War II troopship in 1942 when, with 15,000 American soldiers aboard, she encountered a winter gale 700 miles off the coast of Scotland. The seas were quite large but also quite manageable for the huge ship. Suddenly 'one freak mountainous wave' slammed broadside into the *Mary*, and she 'listed until her upper decks were awash, and those who had sailed in her since she first took to sea were convinced she would never right herself.' After hanging on the brink of capsizing for a few eternal seconds, the great ship finally righted herself again.

"More recently off Greenland, the *Mary*'s sister ship, *Queen Elizabeth*, took a wave over the bow that was so large it flooded the bridge 90' above the waterline.

"In 1966, 800 miles off New York, the Italian liner *Michelangelo* plunged into a gigantic trough that was followed by a huge solitary wave that crumpled the flare of the ship's bow and broke out the inch-thick glass in the bridge windows some 80 feet above the waterline, injuring hundreds of passengers (and killing three).

"In July 1976, the tanker *Cretan Star*, loaded with 29,000 tons of light crude oil, was struck by a huge wave and sunk in the Indian Ocean not far from Bombay. An inquiry reported that the southwest monsoon reaches its greatest strength in July off Bombay and periodically piles up 'episodic waves of vast proportions.'"

In exploring the probability of the occurrence of single large waves, Dr. Lawrence Draper of the National Institute of Oceanography in England used the "Statistics of a Stationary

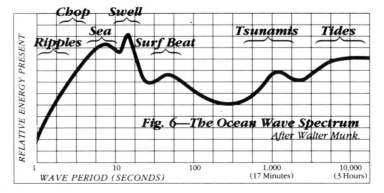

Fig. 6—The Ocean Wave Spectrum
After Walter Munk.

*Lighter than cork
I danced on waves
in the salt air—*

*Waves,
those eternal victim-
tossers, so to say...*

—Arthur Rimbaud

Random Process" to show that one wave in 23 is over twice the height of the average wave, one in 1,175 is over three times the average height, and one in 300,000 is more than four times the average wave height. Put these statistics to work along a stretch of water known as South Africa's "Wild Coast" and you might expect calamity aplenty, and, indeed, that's what you get.

One characteristic of waves is that a following current increases wave lengths and decreases wave heights, while an opposing current has the opposite effect, decreasing the length and increasing the height, thus also steepening the face of the wave. A strong opposing current may well cause the waves to break, even in deep water. Off that southeastern coast of Africa, where the continental shelf abruptly drops away, the Agulhas Current sweeps in hard against this immovable barrier, concentrating the massive southwest flow of water into a relatively narrow stream. The current moves at four to six knots, providing a fast, economical shipping lane for ships moving south. However, when storms to the southwest pump waves around the Cape of Good Hope and up into the channel, the wave length of the swell can be shortened dramatically and the wave steepness increased to precipitous angles. Under certain conditions the unusually swift current here actually doubles the height of the waves pushing upstream. The giants that are created are called "Cape Rollers," and when the statistically predictable rogue wave moves into the current, the result can be catastrophic.

To make matters even worse, waves are refracted, or bent, toward the higher-velocity current, concentrating more wave energy over the strongest current, and possibly even trapping waves there. As Bascom points out, when a ship moving in the current at 18 knots (nine meters per second), assisted by the current of four meters per second, encounters a wave moving at ten meters per second, the velocity of the collision is the sum of these, or 23 meters per second. Since the force of impact is proportional to the square of the velocity, the current nearly doubles that force. "If," says Bascom, "the wave is twice as high as an ordinary storm wave, a ship is likely to be in trouble."

Untold vessels have been lost off the Wild Coast over the centuries. One account from Bascom of a ship that survived conveys the essence of the situation: In December 1969, the middle of the southern summer, the 102,000-ton Swedish tanker *Artemis* ran through a storm on its way down the Wild Coast. Captain L.J. Tarp reported that one wave came over the ship's bow and continued rolling down the deck at such height that it hit and flooded the wheelhouse five decks up.

The complex network of wave relationships—combining, dampening, crossing, overtaking—is a continual dy-

Reef Break Wave
Banzai Pipeline, Hawaii

Offshore winds contribute to the cylindrical shape of these hollow, plunging waves—hollow enough for surfers to ride inside.

An outer "cloudbreak" reef focuses the wave energy and magnifies the size and power of the surf.

On the North Shore of Oahu, the absence of a continental shelf causes the waves to strike the island's reefs with most of their open-ocean power intact.

OVERVIEW

Waves break hard on a fixed reef of rock-solid lava heads. Some of the pockets in the reef are filled with sand, others are swept clean (and made even more treacherous) by the violent wave action.

The offshore flow of water escapes seaward through channels through the reef or around it.

The reef is angled to the approaching swells, causing the waves to peel off in symmetrical precision.

namic of the ocean surface. It is part of what has made the ocean an enduring frontier and a mystery. Always, far beyond the horizon, new storms pinwheel into being, urging up new waves, new swells, out from the otherwise vast, implacable face of the world's oceans.

Most of us will be completely unaware of their distant arising, their silent passage across the empty miles. Most of us will only become aware of them as they emerge out of the distance, touch bottom, rise and finally burst into white glory. Only then, as breaking waves, will their full potential be revealed to us.

Breaking Waves

When long, fast, smooth open-ocean swells move into shallower water, their character begins to undergo a significant transformation. When the depth of the ocean becomes less than half the length between the crests of two successive waves, the speed of a wave is no longer governed by its length, but by the depth of the water; the speed of a wave is now proportional to the square root of the depth of the water it is moving through. It is at this point that ocean swell changes to ground swell. This is where the study of deepwater waves ends and the study of shallow-water waves begins. This is the transition zone between swell and breaking waves.

When swell moves into water less than half its wave length deep, the wave begins to "feel" and be affected by the bottom. The contours of the basin within which the wave travels begin to modify the wave's behavior through a process called "refraction." Refraction here refers to the result of the slowing of waves as they move into shallowing water. This results in a bending of the wave fronts to align themselves to the contours of the shoaling bottom.

Because the speed of a wave in shallow water is a function of the depth, swell refracts as it responds to submarine contours. Since waves slow as the bottom shoals, swells moving laterally toward a sloping coast are bent toward the shore. Similarly, wave energy converges and focuses over shallow ridges [Fig. 8a, 8b], while it diverges and disperses over deeper submarine trenches.

Blair Kinsman reminds us that, "the only feature of a wave as we see it from the beach that has been left unaltered from its deep-water state is the period. You can't tell what direction the waves are running offshore from the angle at which they approach the beach." In fact, as waves move into increasingly shoal water, they begin to slow, the wave length shortens, and the low, sloping mounds begin to rise up out of themselves.

Certainly some amount of drag is produced by the interaction of wave energy with a shoaling bottom; some energy is certain to be released in this way. However, the popular belief that friction slows waves down in shallow water while the crests continue to move more rapidly, and so trip over themselves, seems less popular today than in the past.

Surfer-author-musician John Kelly, Jr. of Hawaii ascribes the change in speed—and in wave height—to deflection, to the idea that ocean wave energy obeys the same laws that control the deflection or reflection of light waves, and that the angle of reflection equals the angle of incidence.

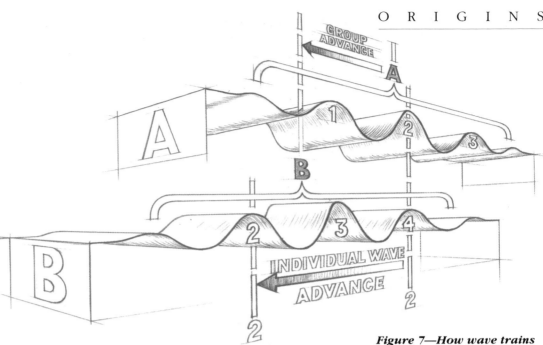

Figure 7—How wave trains travel: *Waves die out over distance and are replaced by following waves at such a rate that the group advances at just half the speed of the individual waves.*

In *Surf and Sea* Kelly writes: "Since the ocean bottom is a fixed boundary, the deflected wave energy is focused upward to a degree that depends on the angle of the rising bottom and appears at the other, flexible boundary of the medium in the form of the rising crest—in effect, an inversion of the wave's energy. As more and more of the wave energy is deflected upward by the confining space of shoaling water, the crest mounts proportionately higher. Here we find an explanation for the slowing of the wave: It is due to the fact that the wave energy, bouncing, as it were, off the bottom and being deflected to the crest, travels a greater distance. The detour consumes time, thus slowing the advance of the wave form even though the energy itself continues to travel in its watery medium at a constant speed."

Although this description might impress some oceanographers as a mere flight of fancy, it does portray a clear (if untrue) image of the dynamics that lead to the breaking of the ocean wave.

As was said earlier, waves in deep water will begin to break when their height is greater than a seventh of their wave length. The maximum stable profile angle of the crest of a wave is, therefore, about 120 degrees. Steeper than this, the wave character begins its final dramatic transformation. In very shallow water, when waves break as they approach land, they will reach this critical angle in a water depth of about 1.3 times the wave height. In other words, a three-foot-high wave will break in approximately four feet of water. It is as waves approach this "limit of their containment" that the most dramatic moments of their lives are played out in the surf zone.

Whatever the lawful causes—friction or deflection—as waves encounter the rapidly shoaling water associated with most beaches, they are said to peak up. That is, their height increases rapidly. At the same time the shallow water causes the wave length to decrease (because as a wave is slowing, the waves behind are catching up); the result is a suddenly steepened wave Therefore, in a very short distance, the crest

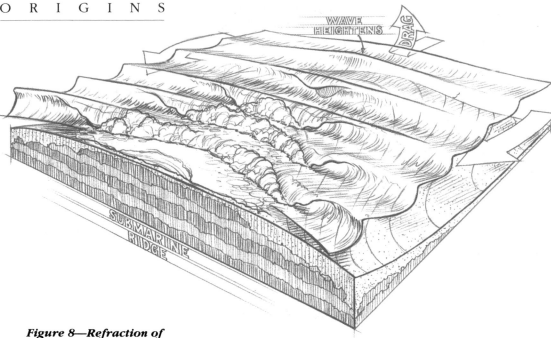

Figure 8—Refraction of waves over a submarine ridge: *The bending of waves as they slow in shallowing water focuses their energy over the shoal area.*

Figure 9—Refraction of waves over a submarine canyon: *The bending of waves as they slow in shallowing water disperses their energy away from the deep water and toward the shoals.*

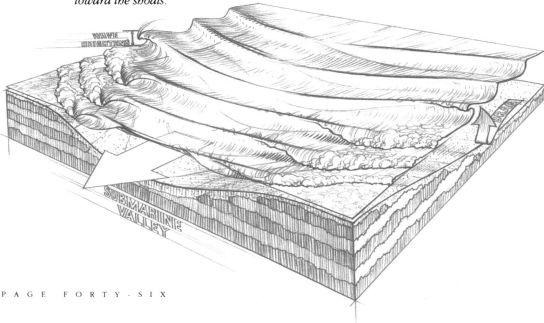

angle decreases below the critical 120 degrees and the wave becomes unstable. The crest, moving more rapidly than the water below, falls forward and the wave form collapses into turbulent confusion, which uses up most of the wave's energy.

Perhaps the leading popular authority on ocean wave phenomena is Willard Bascom. Some thoughts from his *Waves and Beaches* on the dynamics of breaking waves:

"As the swell moves into very shallow water, it is traveling at a speed of 15 to 20 miles an hour, and the changes in its character over the final few dozen yards to shore come very rapidly.

"In the approach to shore, the drag of the bottom causes the phenomenon of refraction…and one of its effects is to shorten the wave length. As length decreases, wave steepness increases, tending to make the waves less stable.

Moreover, as a wave crest moves into water whose depth is about twice the wave height, another effect is observed which further increases wave steepness. The crest 'peaks up.' That is, the rounded crest that is identified with swell is transformed into a higher, more pointed mass of water with steeper flanks. As the depth of water continues to decrease, the circular orbits (the movement of a particle of water within the wave) are squeezed into a tilted ellipse and the orbital velocity at the crest increases with the increasing wave height.

"This sequence of changes in wave length and steepness is the prelude to breaking. Finally, at a depth of water roughly equal to 1.3 times the wave height, the wave becomes unstable. This happens when not enough water is available in the shallow water ahead to fill in the crest and complete a symmetrical wave form. The top of the onrushing crest becomes unsupported and it collapses, falling in uncompleted orbits. The wave has broken; the result is surf."

The energy released in a breaking wave is tremendous. All of that stored wind power—transported silently for so many miles—at last bursts out of its liquid confines with a thunderous roar of liberation. The total energy of a wave ten feet high and 500 feet long can be as high as 400,000 pounds per linear foot of its crest. The impact pressure of such a breaking wave can vary from 250 to as much as 1,150 pounds per square foot. Larger waves have been recorded to exert a force of more than three tons—6,000 pounds of pressure—per square foot in the surf zone!

Echoing the combined energies of the many forces out to sea, ocean waves approach the shore in irregular patterns—cycles of smaller waves and larger waves created by the reinforcing or cancelling interaction of different wave trains. Groups of bigger waves are called sets; long intervals between sets are called lulls. The pattern of sets and lulls—the surf beat—is the pronounced rhythm of the ocean's language, the cadence of its voice.

Waves and Surf

In general, there are three forms of breaking waves: surging breakers, spilling breakers and plunging breakers *[Figs. 9a, 9b, 9c]*.

Surging waves are associated with relatively deep-water approaches to steep beaches. The incoming wave peaks up, but surges onto the beach without spilling or breaking.

Spilling waves are generally produced by a very gradually sloping underwater configuration. The wave peaks up, the crest angle shrinks to less than 120 degrees, but the release of energy from the wave is relatively slow. Spilling waves typically have concave surfaces on both front and back sides.

Plunging breakers are the most dynamic, exciting manifestations of wave action on the ocean. Their rounded backs and concave, hollowing fronts result where an abrupt shoaling of the bottom creates a sudden deficiency of water ahead of the waves, which can be moving at near open-ocean velocity; water in the trough rushes seaward with great force to fill the cavity in the oncoming wave. When there is insufficient water to complete the wave form, the water in the crest, attempting to complete its orbit, is hurled ahead of its steep forward side, landing in the shallow trough. The curling mass of water (called a "tube" by surfers) surrounds a volume of air, often trapping and compressing it. When the trapped air breaks through the curtain of water that surrounds it, there is often a geyser-like burst of spray and mist. Often, too, the mist is expelled out of the open end of a well-defined tube, like smoke from the barrel of a gun. Riding ahead of such a blast of vapor is where a lot of surfers would like to be.

Surfers, animals (including porpoises and seals) and boats are able to ride waves due to the resultant of three forces—the total weight of the vehicle (i.e., surfer and surfboard), the total buoyancy of the vehicle (including planing force), and the "slope drag" created by the angle of the wave's face. When this slope drag is greater than the hydrodynamic drag (water resistance), the vehicle moves at the approximate speed of the wave crest.

One of the major skills required for a surfer is getting the surfboard moving fast enough at an angle precise enough so that the slope drag takes over the work of propelling the vehicle just as the wave rises up beneath him. Once he's up and riding, the surfer can move considerably faster than wave-crest speed by maneuvering sideways across the face of the wave.

Although humans are the most common surfers today,

the act of riding waves is an ancient custom for porpoises, seals, sharks, killer whales, and fish and birds of all kinds. Seals and porpoises are terrific surfers; their instinctive familiarity with the liquid medium allows them to be the most subtle and eloquent wave riders of all.

Because porpoises and seals have neutral buoyancy, they are able to tilt themselves to the correct slope angles of underwater constant-pressure surfaces (literally wave planes within wave planes) and catch the waves there, so that often they are seen in a subsurface mode, imbedded in the wave face as they surf across a transparent wall of water. However, these creatures are also able to break through the plane of the wave face and surf on the outside surface of the wave in a more conventional manner. Interestingly, some human bodysurfers have also learned the art of surfing the underwater constant-pressure surfaces.

Just as not all beaches and bottom configurations help to develop plunging waves, not all beaches—relatively few, in fact—are conducive to the creation of waves for surfing. A perfect wave for surfing is one that is refracted in such a way as to concentrate its power in a given area of the wave band, then "peels" off laterally over a relatively abrupt, shallow bottom so that the wave is a plunging breaker with an extremely concave or "hollow" face. When the crest of such a wave pitches out toward the trough, it can then complete a tunnel-like formation, creating the ideal "barrels" that the best surfers travel the world to find.

Do not move
Let the winds speak
that is paradise.

—*Ezra Pound*

Waves radiating in from the open sea encounter the living reefs that ring Tavarua Island.

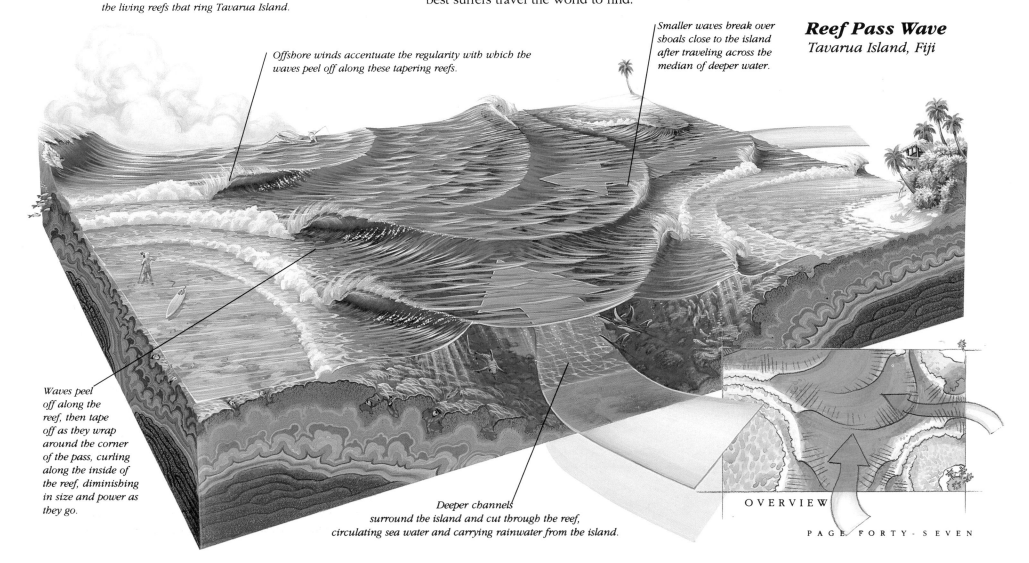

Offshore winds accentuate the regularity with which the waves peel off along these tapering reefs.

Smaller waves break over shoals close to the island after traveling across the median of deeper water.

Reef Pass Wave
Tavarua Island, Fiji

Waves peel off along the reef, then tape off as they wrap around the corner of the pass, curling along the inside of the reef, diminishing in size and power as they go.

Deeper channels surround the island and cut through the reef, circulating sea water and carrying rainwater from the island.

OVERVIEW

In the dramatic last seconds before lines of swell become breakers, waves respond to the sudden shallowing of the water depth by gaining considerably in height (sometimes double or more the swell height), developing a critical concave face, and assuming a beach-facing profile that reflects the immediate characteristics of the bottom shape directly under the wave. These local bottom configurations determine the final form of the breaking waves. In general, there are several types of wave "breaks."

Relatively straight sandy or gravelly beaches with a gentle slope create "beach break" waves—a pattern of peaking waves with periodic channels to carry the advancing water back out through the surf zone. Such waves break on sandbars or "gravelbars," deposits of material mobile enough to be arranged and rearranged at the whim of swell, tide and wind. Often the beach face is scalloped in a regular pattern of "cusps" reflecting the regularity of the coastline, the subsequent regularity of the refraction that concentrates and disperses the wave energy, and the mathematical relationship between the advancing force of waves and the receding flow of water.

Outgoing currents of water between areas of breaking waves are called "rips" or riptides; in large surf they are capable of becoming overwhelmingly powerful channels, moving rivers of water heading back out to sea—frightening locations for swimmers, but ideal onramps for surfers wishing to make it through the near-shore "beach break" waves to catch rides in deeper water.

Very steep sand or gravel beaches are likely to produce surging breakers, where the depth immediately offshore is insufficient to greatly diminish the potential energy in the breaking waves. Thus, most of the wave energy is released directly up onto the beach face or reflects back at the incoming waves; the outgoing sheet of water creates a "backwash" effect that can double or triple the size of an approaching wave, often with spectacular effect.

Submarine formations like coral reefs, rock reefs, sunken ships and other relatively abrupt submerged or partially-submerged formations create "reef" surf—waves that break more or less abruptly and in a variety of shapes, depending on the configuration, depth and size of the obstacle.

The famous Banzai Pipeline off Oahu's North Shore is a reef break; swells radiating in from great Pacific storms to the northwest come out of very deep water to touch coral reefs more than a mile off shore. This outside reef bends the waves, focusing them in on the near-shore Banzai reef with little loss of energy. The waves rise steeply over the "outside" reef, then appear to almost disappear in the intermediate deep-water zone, then abruptly "jack up" as they rush in at the abrupt "inside" reef. There, these giants that have come so far are forced up out of themselves by the sudden wall of battered coral. Immediately there is insufficient water in the trough of the wave for the circulation of water. The

A stream has deposited large rocks, small stones and sand near the tip of the headland. The littoral current created by winter storms has distributed this sediment, from coarse to fine, along the point. The beach farthest inside is sand.

Point Break Wave
Rincon Point, California

Trains of swell marching down the coast refract around the Rincon headland and peel along the cobblestoned beach.

As waves wrap around the point, breaking in the shallows close to shore, their lines radiate out into deeper water

A submarine fan of aggregate stone fans out from the point, refracting waves toward the headland.

The current sweeping down the point escapes through a channel that runs around the shoulders of the waves.

OVERVIEW

face goes as concave as a storm pipe (hence the pipeline name), the crest becomes a "lip" of plunging water that leaps beachward to complete the cylindrical shape of the wave, creating the spectacular hollow within, followed by the familiar blast of mist as the wave collapses around the pocket of trapped air. Add to these fundamental dynamics the angled seaward face of the reef, which causes the wave to peel off to the left and (sometimes) to the right, and it would be hard to imagine a more perfect wave. And, if all this weren't enough, the prevailing wind is off the land and straight up the left-breaking faces of the waves. This has the effect of smoothing the faces, holding them up longer, and allowing them to grow even hollower before breaking.

It is this "shoulder" effect, where the wave can peel along the angled edge of a shallow reef, that makes a wave

of interest to surfers. A triangular-shaped reef, with its apex pointing to sea, will tend to create an initial peak wave, then waves will peel off in either direction as the lines of swell refract and converge alongside the reef. The result can be symmetrical, twisting cylinders of great beauty and finesse.

Assuming a perfect equilateral triangle with its base parallel to shore, such a reef would create the most perfectly balanced peeling waves when the lines of swell approached it squarely. Should the swell direction be from one side or the other, the wave on the near side of the triangle will tend to peel off too fast (to "close out" or "section" ahead of the rider), while the wave on the far side will be "mushier" and less hollow—more of a spilling breaker—as it wraps around the apex of the triangle and disperses its energy into the deeper water beyond.

Peeling waves can also be created by "passes" in coral reefs—channels created in the living formations by the runoff of fresh water from the tropical island land masses which these living reefs tend to surround. Here, the typical surfable wave will wind off the end of one shallow shoulder of reef, peeling toward a deep channel. Such waves can be a mile or more out from shore, and because the reef itself is usually submerged, these walls of water have an isolated and unpredictable beauty.

The one-directional peeling of the reef-pass wave is similar to the peeling of a typical "point" wave, created when lines of swell wrap around a coastal promontory or protrusion and break—often with remarkable regularity and evenness—as they refract around the bend at a relatively constant distance from the curved shoreline.

One fine example is the wave at California's Rincon Point near Santa Barbara—a beautiful triangle of cobblestoned shoreline extending roughly a half-mile from the inside cove to the apex of the point where a creek spills out into the Pacific.

The machine-like regularity with which the swells fan around the point and trace the even shape of the shallow bottom with a surging carpet of white water is a moving impression, to the surfer and wave watcher alike.

Whereas the exposed tip of such a point or promontory will generally have a rocky, gravelly or boulder-strewn beach, the bay into which point waves peel is typically a repository of fine sand. This is because, once a wave has broken into a tumbling chaos of foam, it has lost its internal oscillatory motion. Instead, the particles of water are actually driven forward by the momentum of the wave action. This movement of water toward the beach translates into a strong beachward current along points and promontories. It is this current that is able to move large amounts of sand and other fine particles along the point and into the bay. As energy, momentum and wave speed dissipate, the sand drops to the bottom or washes ashore. For this reason, peeling point waves will often end in an abrupt beachbreak "close-out," as a long section of wave suddenly shoals over a sandbar or surges up onto a straight beach.

Whatever their form, whatever their size, whatever their cause, waves are relentless proof of the power of great outside forces over our lives—the same forces that suspend our world in space, in an intricate web of physical laws—the same forces that sustain the very fabric of our reality.

Waves are carriers of a very important message: that we are not alone, that we are part of a larger whole, and that we are an important enough part of the whole to deserve this lavishly beautiful and magnificent planet. Waves are living proof that something in nature and the universe has quite a high opinion of our intelligence and our capacity for appreciation.

Poetically, waves may be the lips of the sea, eternally communicating—simultaneously—to an infinite number of landfalls. Scientifically and physically, waves are great translators or transformers of energy. If we owe all life to the sun, then ocean waves are, literally, messengers of the gods.

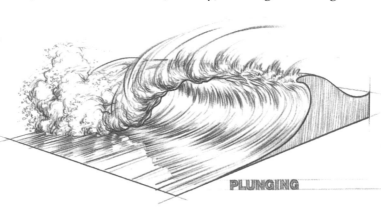

Figure 10—Types of waves: *Depending on local bottom conditions and the steepness of the beach, waves break in three distinct fashions—surging, spilling and plunging.*

BIBLIOGRAPHY

Willard Bascom, **Waves and Beaches: The Dynamics of the Ocean Surface**, Anchor Books, Garden City, N.Y.,1980.

Henry B. Bigelow & W.T. Edmondson, **Wind Waves at Sea, Breakers and Surf,** U.S. Navy, Hydrographic Office, Pub. No. 602, Washington, D.C., U.S. Govn't. Printing Office, 1947.

Nathaniel Bowditch, **Waves, Wind and Weather,** David McKay Co., N.Y., 1977.

Rachel L. Carson, **The Sea Around Us,** Oxford University Press, N.Y., 1951.

Vaughan Cornish, **Waves of the Sea (and Other Water Waves),** Open Court Publishing Co., Chicago, 1910.

C.F. Hickling & Peter Lancaster Brown, **The Seas and Oceans,** MacMillan Publishing Co., N.Y., 1974.

John M. Kelly, Jr., **Surf and Sea,** A.S. Barnes and Co., N.Y., 1965.

Cuchlaine A.M. King, **Beaches and Coasts,** Edward Arnold Ltd., London, 1959.

Blair Kinsman, **Wind Waves,** Prentice Hall, Inc., Englewood Cliffs, New Jersey, 1965.

Pierson, Neumann & James, **Observing and Forecasting Ocean Waves,** U.S. Dept. of the Navy, Washington, D.C., 1955.

R.A.R. Tricker, **Bores, Breakers, Waves and Wakes,** American Elsevier Pub. Co., N.Y, 1965.

Encounters

And I have loved thee, Ocean! and my joy
Of youthful sports was on thy breast to be
Borne, like thy bubbles, onward; from a boy
I wantoned with thy breakers,—they to me
Were a delight; and if the freshening sea
Made them a terror, 'twas a pleasing fear,
For I was as it were a child of thee,
And trusted to thy billows far and near,
And laid my hand upon thy mane
   —as I do here.

—Lord Byron

XX

XXI

XXII

XXIII

XXIV

XXV

XXVI

XXVII

FACE TO FACE WITH SIGNIFICANT WAVES

Pleasant it is,
when winds disturb
the surface
of the vast sea,
to watch from land
another's mighty
struggle.

—Lucretius

While most of us are content to enjoy breaking waves from a distance—though some might even adventure within the great laws and theories of physics and mathematics to study them—others choose (or are forced) to experience the sea and the surf directly. For many of these, the great encounters of their lives will be with ocean waves.

Historically, the ocean has been an obstacle to the aims of men; waves have been the sea's sinister messengers of doom, laying in wait to dash the dreams of the hapless, administering a cold kind of justice upon those whose only sin was insignificance.

Imagine the fearful, storm-tossed months and years at sea of the Greeks who sailed with Odysseus. What terrible waves roared over the fragile ships of Leif Ericson or Columbus or Magellan? What great walls of water are etched into the memories of those who have been to sea and into the final fates of those who have died there? What murderous waves have swept suddenly out of the night to claim whole societies of coastal dwellers?

Legend has it that a man called Holua was swept off the "Big Island" of Hawaii (along with his house and most of his village) by a seismic ocean wave. In the raging confusion out at sea, Holua grabbed hold of a large plank, crawled aboard and stroked desperately into the next great wave, surfing the 100-foot monster into the beach and safety—the largest and longest ride in surfing history. What a tale to tell his family…if there were family left to tell.

This ancient foe—the ocean—remains as wild and free today as in those ancient times. It is still a great proving ground for the spirit of man, a medium for essential rights of passage—a place to step outside the laws and test oneself against the objective standard of implacable nature.

Ed Perry went truant from school at 14, hopped a freighter to China and began a life at sea. Although he lifeguarded summers at Venice, California, he logged an impressive number of voyages in his 58-year career, first as a hooker on a whaler, then in the merchant marine, and finally (for 38 years) as a captain on tankers hauling oil out of the Persian Gulf. In all those years, Perry best remembers the storm of 1933 and the great waves that were piled into monsters by a huge typhoon whirling in the South Pacific.

Perry was a deckboy aboard the *Jutlandia*, a 7,000-ton, 428-foot-long vessel originally built for the King of Denmark, then sold to the East Asiatic Company. The *Jutlandia* was one of the first big diesel yachts (it sported a familiar steamship stack only as a cosmetic nod to the romantic ideas of prospective travelers), but it was never designed to carry passengers or cargo on 'round-the-world voyages.

The ship took on a big load of lumber at Grays Harbor, Washington. Finished wood was stored below; tons of rough planks were lashed to the decks with chain. The Captain selected a great circle route across the North Pacific, and the passage was uneventful until, east of the date line and south of Japan (at about 140 degrees east and 30 degrees north), the *Jutlandia* began to encounter the advancing swells from a major storm system spinning up from the South Pacific.

"It started getting rough," Perry recalls, "and then it got rougher and rougher and rougher." As the storm was spinning, it was also moving in a tight arc, veering toward the northeast. Because of the tightness of the arc, the winds were creating successive trains of swell that converged at different angles to the north, causing the waves to wedge up into larger, steeper peaks than would be expected.

Only gradually did Perry begin to sense how bad the situation was becoming. "I figured that the officers and

Captain knew what they were doing," he recalls (he was 19 at the time), "until one morning I found out we weren't going to have any breakfast because we'd taken some seas up forward and it had broken some of the chain lashings. We were losing lumber over the side real fast. In fact, a couple of huge waves picked up some of the lumber—12x4s about 20-feet long—and threw them back over the bridge, and they came down through the skylights into the galley, which was located mid-ship. So we were getting seas coming down into the galley and going right on into the freezebox. Then some of the lumber broke the skylights above the engine room and we started taking water down there."

The waves were 40- to 50-feet high, not an unheard-of size, but they were very steep. The incoming water was absorbed into numerous bales of rags (stored below for cleaning purposes); the rags burst out of their ropes and floated free, clogging the rose boxes and knocking out the ship's pumps. Eventually the Captain had the entire crew and all passengers turn-to to bail the *Jutlandia* with buckets. Life jackets were ordered worn 24 hours a day. The ship made Shanghai 19 days late sporting a 17-degree list with not a stick of lumber on deck. Perry could count himself among the lucky survivors of the "Storm of '33," and the memory would still burn bright over 50 years later.

Certainly the morning of February 7, 1933, must have long after burned bright in the mind of Lieutenant (j.g.) F.C. Margraff. He was the watch officer on the bridge of the Navy tanker *Ramapo* as it toiled through the peak of that same seven-day storm and its enormous waves. With near 70-knot winds directly astern and being chased by mountainous following seas, Margraff observed the highest wave ever reliably estimated—at least 112 feet! What other experience in a man's life could compare to a mid-Pacific encounter with a wall of water taller than a ten-story building?

The ocean is an ever-changing frontier, a vast and dynamic wilderness swept by great natural forces. Yet, as alien a place as it might seem for clumsy, air-breathing creatures like men, the sea has been the great provider, a source of livelihood, a medium for material and cultural exchange. And, almost from the beginning of history, this great watery desert has also been used by man as a medium for sport, for personal testing, for accomplishing those vital rites of passage.

The Polynesians have ridden breaking waves for hundreds of years. Who knows who rode them before? Who knows but that surfriding might have been the noblest recreation in all the island state of Atlantis?

We do know that surfboarding, after capturing the poetic imaginations of Captain Cook and his immediate followers, almost died out in Hawaii in the mid-1800s, was resurrected when several inspired visitors (like Mark Twain, Tom Blake and Jack London) validated it with their appreciation and attention, and finally began to flourish in Hawaii, California and Australia in the 1920s and '30s.

Almost from the beginning of this surfboarding renaissance, the most captivating challenge was riding big waves—the "blue birds," the "cloud breaks," the giants that broke far from shore with Olympian grandeur. And, in a real way, the men who were able to ride them were accorded a kind of god-like respect within the society of their peers.

The lure of big breaking waves has now attracted generation after generation of surfers—bodysurfers and board surfers alike. Even paddlers have taken up the gauntlet: on New Year's day, 1980, three Hawaiian men (Aka Hemmings, Dale Hope and Tommy Holmes) paddled their 22-foot canoe, *The Duke*, into a 30-foot peak at a place called Avalanche off Oahu's North Shore. There were some injuries, but all survived to try again.

For decades surfers have searched for the ultimate wave, wondering how big was too big. Waimea Bay and Makaha, both on Oahu, are two fabled big-wave spots, both

The power and the glory: Big-wave pioneer Greg Noll contemplates the magnitude of the odds as he looks out from Banzai Beach on the North Shore of Oahu at the savage winter waves of the Pipeline Photo circa 1962 by John Severson.

triggered by reefs lying off of prominent headlands. Each has its great days, and when those days happen a few great surfers come out to meet the challenge. Sometimes the deeds they do are almost beyond belief, and beyond survivability.

When Mark Foo challenged Waimea Bay on January 18, 1986, it was in the best tradition of accepted North Shore big-wave assaults (a tradition that includes not only being rescued by a Honolulu County helicopter, but actually being dropped from a helicopter beyond the surf zone when there is no possibility of paddling out). Not in the tradition of big-wave assaults on that particular day was the rogue set that poured through—a 50-foot wave, witnesses say.

"I paddled over a solid 25-foot wave," Foo said later, "and I saw this thing…the biggest thing I'd ever seen. The funny thing is, I think I laughed…it was so unbelievable. It was like a cartoon almost." Foo survived; the helicopter brought him home.

But not all great, mind-altering wave experiences are of the life-or-death variety. Listen to Fred Van Dyke, a man who rode giants for over 25 years. It was the early '50s in the cold-water paradise of Santa Cruz, California, and Fred was a teacher at the time, making $3,000 a year to support his surfing habit. He'd just made the 15-minute drive down from school in the San Lorenzo Valley, then stopped at the Coast Creamery on Mission for a milkshake or two while the wind died and the ocean got glassy before dark. Listen to Fred tell about one wave:

"It was big Steamer Lane. It was late in the evening. We didn't go out in the water until about 4:30, and in the winter it got dark about 5:15 or 5:30 at the latest. This particular afternoon, after we'd had our milkshakes to fortify ourselves for the 46-degree water, we jumped off the cliff at Steamer (which you don't have to do now)—you'd drop your board, jump in as close to it as possible, swim to it, get on and paddle like hell to get outside, then you'd sit out there and wait. Well, I caught a couple of waves, and it was getting pretty dark. The moon was coming up over the bay to my right, and off to my left the last arc of the sun was setting in the redwoods above the cliffs, so it was a fantastic, incredibly beautiful late afternoon…dusk…just about a full moon…and I wanted to get in. The water and air were

about the same temperature so the water actually felt almost warm. I'm sitting there waiting for the last wave, and everything is totally beautiful. And then it comes. Silently. And I take off on the last wave—it's the outside reef at Steamer—and I make the take-off, and it's dusk, and I don't know where the hell I'm going. I'm afraid I'm gonna hit a piece of kelp, but I make it to the bottom and turn up the face, and as I turn up the face of the wave I notice my friend is bodysurfing behind me, and I let out a scream because that's exciting—to see somebody bodysurfing the wave right behind you—then, right in front of me, a sea lion pops out of the wave and starts surfing, throwing water off its chest, and I hear it going shhhhh—shhhhhhh—shhhh*hhhhh*. And as I surf around the point, making a left turn back toward the cliff, all of a sudden ahead of me on this four- or five-foot wave, a 45-pound salmon jumps completely out of the water, and it's shimmering in the last bits of golden sunlight, but it's also shimmering on the other side from the moon-

The ocean remains a great proving ground for the spirit of man.

Right: *The thunderous Waimea Bay shorebreak on Oahu's North Shore. Photo by Rick Doyle.*

Opposite: *Backwash at the Santa Barbara, California sand-spit creates bizarre rogue oscillations. Photo by Al Daniels.*

Below: *Surfers off the North Shore of Oahu scramble at the approach of an "outside set." Photo by Steve Wilkings.*

light, and I'm so stoked that as I go around that point, I pearl the board purposely into the water and do a back-flip off it, screaming ecstatically. It was such an exciting thing! It cost me a cold swim in after my board, which was carried into the cliff, but it was all too much—just the fact of taking off on a perfect wave at sunset with the full moon coming up, a human being on my left, a sea lion on my right, then the salmon jumping—all that together was an extremely exciting experience. I felt so in touch with things…I experienced it all so fully, I enjoyed it so much that my whole relationship with nature just came together at that point in one great experience. Everything was one."

Other voices tell of other waves, of other peak experiences with the ocean. Looking back through the lens of time at a momentous day 30 years earlier, two surfers each recall waves that changed their lives on perhaps the most harrowing days of their lives.

"I woke after tossing and turning all night long," recalls John Severson, founder of *Surfer Magazine*. It was January 13, 1958, and he was a peacetime soldier stationed at Schofield Barracks on Oahu. "I knew the surf could be giant, so I hitchhiked over to the base at five in the morning and asked for a three-day pass, saying that I was wiped out. The pass was granted, and I hitched a ride in a dump truck going over Kolekole Pass to Waianae. At the top of the pass I was stunned by swell lines stretching to the horizon. Nanakuli Point looked like a monster Malibu; it took the waves three seconds to fall from top to bottom, so I knew it was *really* big.

"By the time I reached my shack in Nanakuli I was jumping. Adrenaline had erased all fatigue. I telephoned Fred Van Dyke on the North Shore to tell him what was happening, but when I got to Makaha it was a disappointing 12 to 15 feet with only George Downing, Wally Froiseth and Buzzy Trent on the scene. It was powerful but not happening, and they suggested we check the Maili cloudbreak [giant waves breaking on primary reefs far offshore]. 'Okay,' I agreed. I was frothing at the mouth for something really big.

"At Maili, it was plumes of spray to the horizon. We watched. I was ready. But George said there was no way to find a lineup [a spot to catch a wave safely], and it seemed to be getting bigger. Maybe Makaha Point had picked up, we speculated, so back we went.

"Makaha was obviously bigger, and I was tired of standing around, so I said, 'Let's go!' George said he wanted to check it out a little longer, but Buzzy said okay and soon we were paddling.

"Buzzy kept indicating possible lineups as we paddled out, but the sets were breaking out farther still. He began raving about how much of the southwest Waianae coast we were now able to see. Then a really big set fringed about a quarter-of-a-mile outside us, and we just shut up and paddled for our lives. We just made it over. Then we sat for awhile, surmising that it had been a freak set, and that we weren't going to catch any waves out here in the middle of the ocean.

"So we edged in, and a nicely-shaped set approached. Buzzy said, 'Let's go!' and we both took off on what I thought was about a 15-foot wave. But the wave jumped as it came in over the reef, and as I swooped off the bottom

and up into the middle of the wall, I could watch Buzzy ahead and way above me as we raced across the longest big wave I'd ever ridden. I looked back and saw a great tube with a pouring lip that looked six-feet thick where it impacted on the flat water.

"But looking back was not comforting, so I focused on the beauty ahead. It was still morning and there was a beautiful sparkle on the wave. It was perfect to the end, and we just eased over the shoulder as the wave moved into the deeper water of the channel. There had been no notorious 'Bowl' [a dangerous wedging section of the wave], for which Makaha was so famous, and I didn't know this was not typical for Makaha 'Point surf.'

"In fact, I had plenty of previous experience with the Bowl, but I figured the reef that caused it was well inside us now, and therefore was not affecting these waves. I still thought the wave we had ridden was not *big* big, but Downing later told me it was about 25 feet.

"We paddled back out, stoked and ready for more, and situated ourselves in about the same spot. Then a bigger set approached. We paddled out over a few waves, then an incredibly large green beauty loomed as an almost perfect peak. This was a big wave—in the 20-foot range, I figured.

"'Let's go, Buzzy!' I said, and we both started to paddle. What I really liked about this wave was the 'peaki-ness' of it. It looked like a good chance to warm up on a big one without facing a likely wipeout and a bad swim. I was riding my new international orange Pat Curren-shaped 10'8" elephant gun [a big-wave board] and paddling like mad, yet I was still not able to feel the energy of the catch. Buzzy was off to the 'shoulder' side of me, and he just couldn't get over the hump.

"I knew I was just about into this big beauty, so I stood up and took a couple of steps forward to push the front of the board down over what was now a ledge of

*Once more
upon the waters!
yet once more!*

*And the waves
bound beneath me
as a steed that
knows his rider.*

—Lord Byron

The floods are risen,
O Lord,
the floods have lifted
up their voice:
the floods lift
up their waves.
The waves of the sea
are mighty,
and rage horribly:
but yet the Lord,
who dweleth on high,
is mightier.

—*Prayer Book,* 1662

water. The wave was jumping fast, and as I started down I back-pedaled into position and raced down the face of a bona fide *big* wave. I hadn't looked around during the last moments of paddling and was now shocked to see that the peak had rippled ahead and I was turning into an incredible wall with a section 50 yards ahead already leaping out into space. This was the famous—and dangerous—Makaha Bowl.

"Man thinks fast when confronted with the potential big *tamale,* and my mind was working: 'Wow, this is beautiful and awesome…sun reflections bouncing off the glassy blue surface…looks like the famous Makaha photo of Jim

Fisher only about twice as big…if I straighten out I'm going to get nailed by the tube, so my only chance is to use my speed and do a pointed dive into the wall and try to get beneath and maybe through the back'…which I did about a half-second after this thought process started.

"I didn't skip on the surface. I speared deep into the wave, and everything went quiet. I'd made it. I remember no rush of water. Only hitting the bottom feet-first, like I'd just jumped out of a second-story window. I'd been swept over the falls in the lip, which then drilled through some 30 feet of water with me caught in the moving lip, and so I felt almost no water resistance. And there I was, squashed into a squatting position on the ocean floor, but still quite alive. I would wait a few seconds for the white water to subside, I thought, then I'd push off and swim to the surface. I felt lucky I hadn't landed head first.

"After a few seconds I decided to push off and test the going. I was shocked to find I couldn't move—not even unbend from the squatting position. Quickly I remembered the old 'rag doll' theory and relaxed to save energy. I held as long as I could until I got that 'I need air quick' feeling. I pushed off and started up, but was slammed back to the bottom. I waited a while longer.

"Things were getting critical, but there was no panic. I was blacking out, but I would not give in to gasping. I came back to consciousness and pushed off again. It was still churning, but I wasn't losing ground. At this point I was really concerned with the next wave. If it rolled over before I got to the surface I was done for sure. 'Well, that's the way it goes.'

"Euphoria swept me. My life flashed by…and the realization that I was paying for high adventure with my life. But it seemed like a fair trade. I thought of my loved ones and said good-bye and blacked out again.

"A half-consciousness came back. The water was

lighter. I was near the surface, but my arms were like lead. I could barely move but managed to struggle in slow motion for the surface and *air!*

"My whooshing suck startled Buzzy, who was sitting not too far outside in the lineup. 'Severson, what are you doing here?' was his first remark. 'I thought you proned-out to shore.' A couple of minutes had passed. I could hardly talk.

"'Bad wipeout…I'm alive.' My ribs hurt. My foot was bleeding. I took off for shore in a slow-motion crawl. That was the end of Point-riding this swell. The waves now folded over in one mighty wall.

"How big was the wave? Over 30 feet, I'm sure. But how *thick?* Although it almost took my life, it gave me a new life—almost like a bonus. I was never comfortable in 20-foot-plus surf after that, but loved the next notch down.

"Later, on shore, I watched George Downing, paddling far outside, get caught inside the biggest wave I've ever seen. I could see he wasn't going to make it and thought he had a good chance of dying. He bailed and dove, making it under, but in the after-turmoil I didn't see him again, and I was anxious until he showed up on the beach later.

"The next day was almost as big, with more surfers. The waves were so fast it was a continuing series of free-falls off the tops of great sucking tubes. 'Made' waves were the stuff of lifetime memories. I know I got mine."

The surfer Severson saw trapped by that giant wave was, perhaps, the only man on Earth who was capable of surviving it. George Downing was an expert surfer and consummate waterman. He spent most of his life in the ocean, swimming, paddling, surfing and bodysurfing. He was always in training and knew exactly his capacities and limitations. In fact, a day earlier he'd found himself at his outermost limits. Downing's voice:

"The Hawaiian Weather Bureau had announced a high-surf warning, and big waves were expected to hit the north and west shores off all islands. So Wally Froiseth [one of the great surfers of the '30s, '40s and '50s] and I decided to go down to the weather station to look at the charts. One glance, and we gave each other one of those looks—this was one of the biggest low-pressure systems we'd ever seen, with wind velocities that were sure to generate enormous surf.

"We decided to pack up our boards that night and head out to Makaha, our favorite big-wave spot on the northern stretch of Oahu's west shore. When we got to Makaha, the waves were no more than a couple of feet high, but we knew what was coming. We parked out by the Point and tried to sleep, but the adrenaline in our bodies was at a full race.

"Wally was older, wiser and more experienced. He kept telling me to get some sleep—we'd need it—but I kept talking, asking him all kinds of questions. Finally I dozed off. The next thing I knew, Wally was shaking me awake. It was still dark, but you could tell dawn wasn't far away.

"'Well, Laddie, it's started to arrive,' he whispered. 'Waves so big and beautiful you won't believe them.'

"Wally's ears had stayed tuned to the surf all night. With his years of experience, he could tell almost exactly the size of the surf by its sound. He had waited to wake me because he wanted me to see no less than 20-foot Makaha 'Point' waves. I never, ever questioned Wally's ears.

"It was now around 5:30, and we walked out toward the end of the Point. The only light came from the stars above. The closer we got to the Point, the louder the surf became, the thicker the salt in the air.

"I'd already had some experience in big Makaha Point surf, so even with my younger, less experienced ears, I knew from the crashing roar that this day, indeed, would be one of the biggest ever. We stood at the Point and waited for light.

"During those minutes, Wally and I didn't say much to each other. I'm sure we were both searching our personal selves, waiting for the light that would show us what the day had in store. I'd grown up with Wally, and there was no pretending between us. We both knew what the odds were on big waves, and size was never an issue. Conditions were the critical factor. We'd both come to surf, regardless of how big it was, as long as the swell direction, wind and chop conditions allowed us to catch a wave. Because, at a certain size, waves become impossible to catch, let alone ride. So the ultimate question became: How big a wave can I paddle fast enough to catch before the white water or the lip pitches out causing my surfboard to lose contact with the wave face?

"Having had some experience in giant Makaha Point surf before, Wally knew the price you paid for a bad wipeout. So when he stood beside me that morning and spoke, I listened carefully. This was a big-wave master talking—my teacher, my friend, my surfing buddy.

"The growing dawn gradually revealed the surf. Without saying a word, we looked at each other just like we had at the weather station. 'This is a perfect day,' Wally said. 'The conditions couldn't be better.' He paused, then added: 'Let's go before the crowds appear.' Wally never lacked a sense of humor, and always at the right time.

"We ran back to Wally's 1936 Black Phantom touring convertible and drove back around the corner to Makaha Beach. As we pulled up, who should be waiting, waxed up and ready to go, but Buzzy Trent. He said he knew it was the two of us standing out on the Point in the first light.

"Buzzy was our good friend and surfing buddy. He was one of the few big-wave riders who understood the conditions that made Makaha good, big and rideable. Buzzy loved Makaha Point surf. He'd surf it no matter how big it got, and he surfed it as well as anybody. Wally and I were stoked that Buzzy was with us to see and surf the Point at its best.

"This day was to go down in each of our minds as something special, a day we would not forget for the rest of our lives. We paddled out together. There was enough time between the sets to make it through the shorebreak easily. We paddled through a small opening in the channel between the waves peeling over from Makaha Point and those peeling from the other side of the bay, from a break called Klausmeyer's. Both breaks were already showing signs of closing out occasionally.

"When we reached the outside, near the area called 'the Bowl,' a big set appeared on the horizon. We decided to sit and watch it so we could establish a take-off lineup.

"It's one thing to visualize a 20-foot-plus wave, but it's an experience of a different order to sit on a surfboard out in

Significant encounters with big waves are the stuff of great drama.

Opposite: *On January 25, 1978, the fishing boat* Mojo, *under charter to actor George C. Scott, was caught by a rogue wave at the entrance to the harbor at Morro Bay, California. Although there were some injuries and major damage to the bridge of the vessel, all survived. Photo by Scott J. Redd.*

Above and left: *On January 1, 1980, three Oahu paddlers set out to ride a "significant" wave in a traditional Hawaiian outrigger canoe. After a nearly disastrous first ride (in which one paddler was injured) the remaining two stroked over the abyss and down the face of a twenty-five-foot giant, chased by rolling thunder at a North Shore spot called Avalanche. Sequence photos by Steve Wilkings.*

Waves
in endless motion
Playful
Forceful
Beautiful
Forgiving
Always satisfying
Like a woman
in disguise.

—George Downing

the channel and watch a set of seven to ten 20- to 30-foot waves break perfectly. We were speechless, watching these beautiful monsters peel off from Point to channel with no Bowl—perfect wave faces with thick lips throwing out to form tubes big enough to drive a bus through and not wet the sides of the body.

"Big-wave experiences end up as trophies in your mind. Waves of the size and quality that we experienced on this day can never be forgotten; they leave a trace, a never-ending energy that fuels the memory, bringing back the images to appreciate again and again.

"Like Buzzy and Wally, I can remember those waves as if it all happened yesterday, from the first decision to paddle out and face the challenge to those moments in the lineup watching that first huge set. I remember seeing Buzzy and Wally paddling for waves that were so big, so vertical, that I had to stop paddling, to stop and yell out, *'No! No! Buzzy! No! No! Wally! Too steep! Don't catch it!'* But my concern for them would vanish as they slipped beneath the pitching lips and got to their feet. The technique of knowing how to trim the board high on a vertical wave face allowed all of us to surf and survive some of the biggest Point waves ever ridden.

"We surfed from about 7:00 a.m. to around noon, and the waves had increased in size each hour we were out. With long swims, heavy balsa boards and many deep dives, the morning had taken its toll on our energies, so we went in to eat. I finished fast and decided to go out again, knowing that Wally and Buzzy would follow.

"I paddled out through the channel and into the same lineup we'd established earlier. I rode two waves, then sat through about a half-hour lull. Then I noticed, way on the western horizon toward Kaena Point, a very unusual, black shadow. At first I thought it was a trick of light, of sun and cloud, then I realized it was a set of waves. This alerted me

somewhat, but I wasn't alarmed by it. I checked my lineup, to make sure I hadn't drifted in too far or too far over toward the Point where the waves could catch me inside them, but I saw I was on target. Still, I decided to move out 50 yards, just to be safe.

"By now the shadows had become what I suspected and had begun to fear—huge, surreal swells that were still a ways up the coast off Yokohama Beach and the Makua Caves. As I paddled out farther, I talked to myself about what a pussy I was for moving so far outside my lineup. 'How could any wave break this far out?' I scoffed at myself. I had paddled so far that I could no longer distinguish cars or people on shore. 'How could a wave be any bigger than what we rode or had seen breaking already?'

"And yet I found myself paddling out faster, driven by a rising emotion of urgency, a prickling impulse for survival. To this day, I cannot forget this experience of logic taking a back seat to instinct.

"By the time the first wave of the set arrived, I had moved out some 200 yards and away from the Point toward the channel another 50 yards. That morning it had been impossible to catch a wave from this far outside the lineup. But when I finally got a clear look at that first giant, it was well outside me still and already starting to develop into a full breaking wave, bigger than anything I'd ever seen before.

"I was far from shore and caught in a bad situation. I was too far out to risk paddling back toward shore—the first waves of the set could catch me right in the impact zone and, if that happened, I'd be lucky to survive. But if I paddled farther out, I risked being caught in the impact area of an even larger wave. I decided to paddle out.

"The first wave was already pretty steep as I stroked up the giant face and broke over the top. The second wave was larger still, already vertical and starting to curl over at the top. It was a long climb up the front of the wave, and I was beginning to dread what might be behind it.

"The third wave was larger still. Its sheer face was dark and glistening, and the water in front of it was being sucked up into the wave, so that I was paddling down into this frightening trough before I started pulling myself up the enormous vertical wall of water. I burst over the crest as the wave hissed by beneath me, only to have my heart sink at the sight of an even bigger wave blocking out the horizon ahead.

"I paddled hard to make it over the fourth wave, angling toward the deep water of the channel, although it was obvious this wave was going to break across the entire bay. It seemed tó take an eternity to make it up and over. I could feel the force as the huge curl crashed behind me.

"In front of me was a wave bigger than all the others. It was already starting to curl over at the top, far above me. Its face was sheer and pock-marked with boils caused by its suction on the reef far below. I had been paddling for a while now, but it was not fatigue that made me stop. It was something else that's hard to put into words.

"I was coming up out of that awful trough, starting up the face, when I looked over to my right, toward the Point, where this most incredible wave was already breaking, pitching out a lip so thick and powerful it was beyond my

comprehension. And the space that it framed, that enormous tube, was so massive, on such a different scale from anything I'd ever seen, that it was just too much. Then I realized that the wave was actually sucking me up the face. I pushed my board away and dove for the bottom.

"I believe to this day that if I had not gotten off my board and clawed my way through the face of this wave, I would have gone over the falls, or for sure, have been swallowed in the hole created by the breaking wave.

"As I penetrated into the base of the wave, I instantaneously was in 30 or 40 feet of water. There was a lot of pressure. Then, as I dove deeper beneath the wave, I was surprised that the water was very clear, and I could see the bottom.

"I felt the pull first, nearly taking me with it, then the concussion and pressure of the wave breaking. Knowing I'd escaped its hold, I turned shoreward, stopping to watch this one great wave break from underwater. I estimated that I was in 50 to 60 feet of water. I could clearly see the impact, some 60 to 80 feet inside me. A pure white upside-down mushroom cloud formed, rolled downward, hit the bottom, then started to burst up and outward toward me. I realized that this wave was indeed something special in size and force as boulders on the bottom tumbled in the turbulence, knocking together with dull, clunking sounds.

"I cut short my fascination with this show to pull myself back up to the surface. Was a bigger wave about to drive me to a violent end on the reef below? I had no idea, and broke the surface prepared for the worst.

"The sea was calm. There was no ultimate giant in sight. The only waves were moving out instead of in. I was amazed. For a wave to open such a massive hole in the ocean, fill it with great volumes of white air and generate refraction waves on the surface heading out to sea like a backwash—all this happening in 50 to 60 feet of water three-quarters-of-a-mile from shore—this was an amazing thing.

"In the excitement, I didn't realize that the sudden increase in depth pressure and the concussion from the breaking wave had me bleeding profusely from the nose.

"Many more sets came through that afternoon, but they were like the waves that morning. I caught a few more rides, but I've never experienced anything like the one I didn't ride."

Waves: terrible and beautiful...endless sources of delight and enlightenment...forms for contemplation or confrontation...infinite play and infinite meaning...and, like all life on Earth and reality itself, miraculous.

Waves—the terrible and the beautiful.

Opposite: *Taking the big drop under a heavy-water cornice at Waimea Bay.*
Photo by Ted Grambeau.

Left: *The lunge of the cobra.*
Photo by Jeff Divine.

Below: *All through the winter, day after day, the spectacle continues at the Banzai Pipeline...forms for contemplation and confrontation.*
Photo by Jeff Divine.

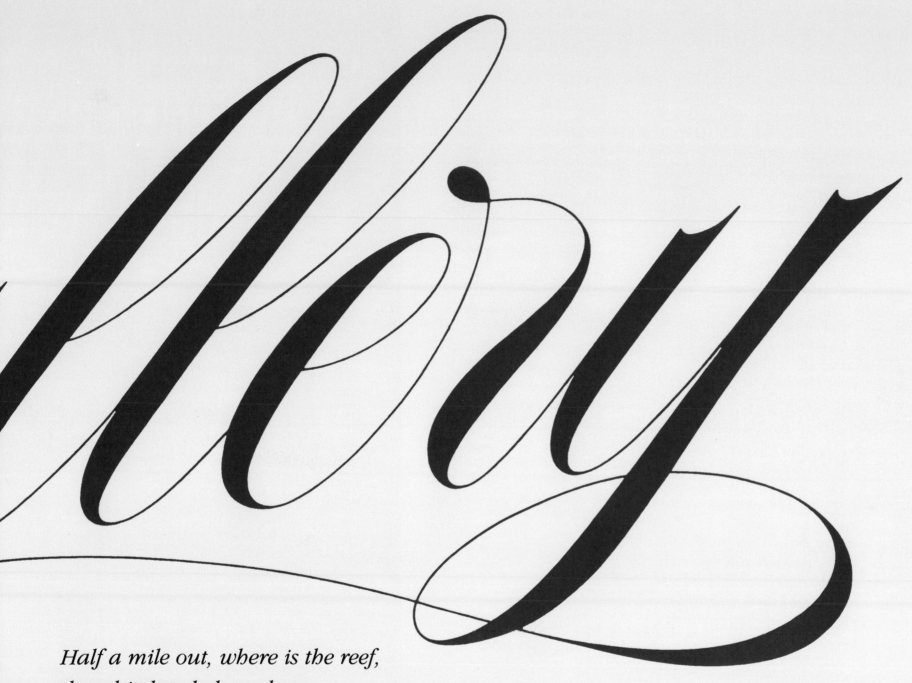

Half a mile out, where is the reef,
the whiteheaded combers
thrust suddenly skyward
out of the placid turquoise-blue
and come rolling in to shore.
One after another they come,
a mile long, with smoking crests,
the white battalions
of the infinite army of the sea.

—Jack London

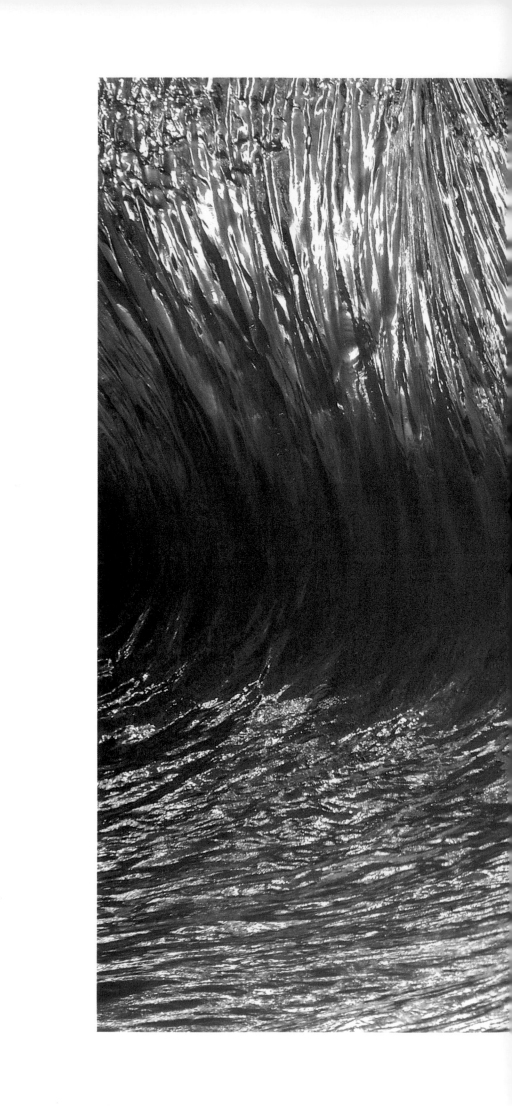

How the water sports and sings!
(surely it is alive!)

—Walt Whitman

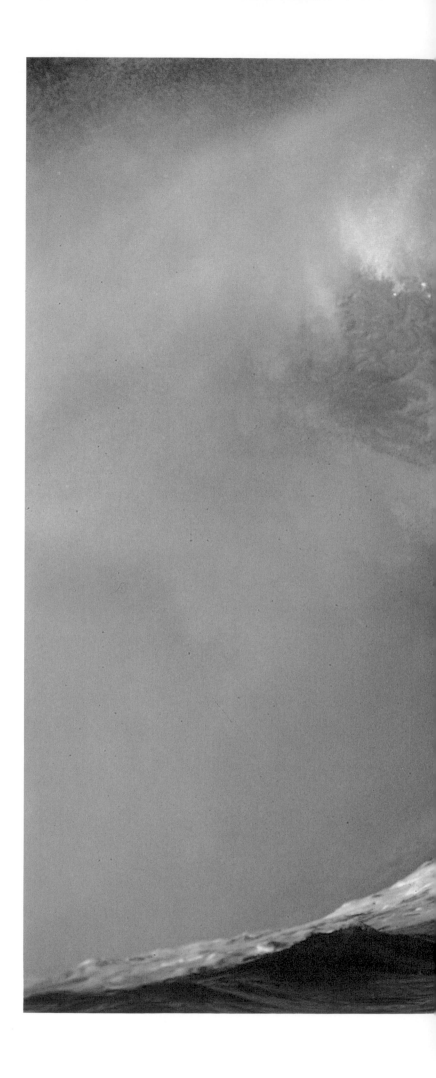

XXX

XXXI

XXXII

XXXIV

XXXV

Like as the waves make towards the pebbled shore,
So do our minutes hasten to their end;
Each changing place with that which goes before,
In sequent toil all forwards do contend.

—William Shakespeare

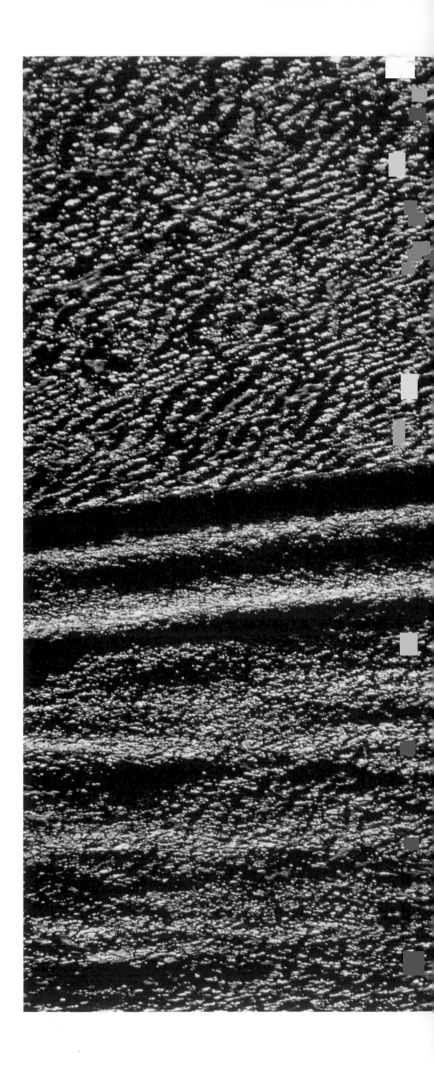

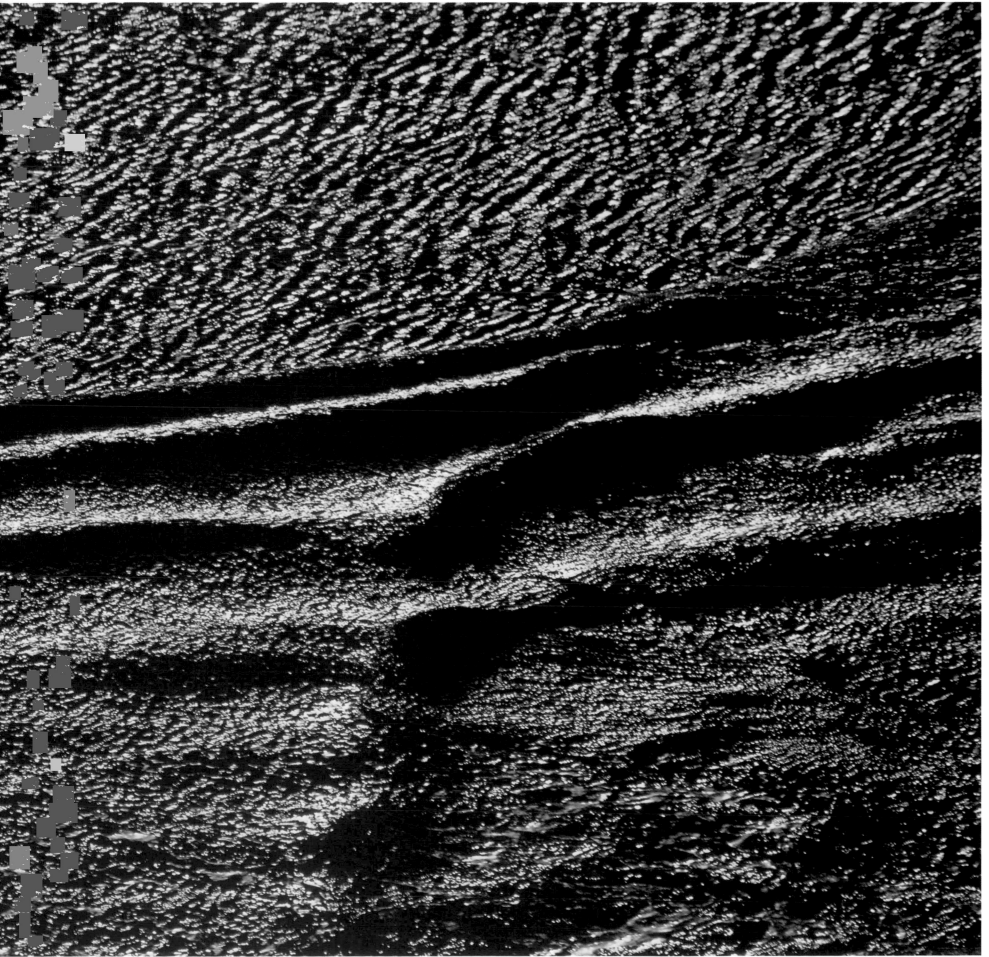

XXXVI

XXXVII

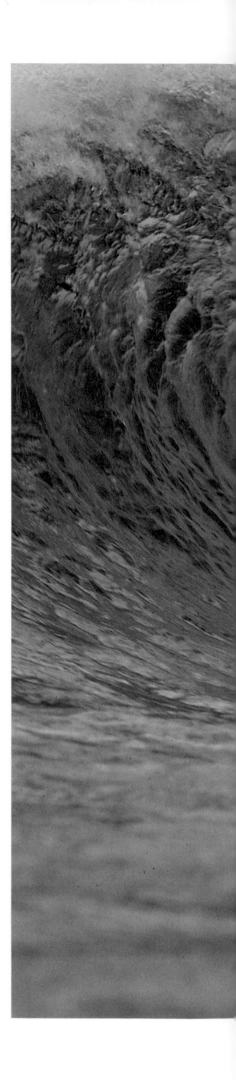

XL

XLII

XLIII

XLIV

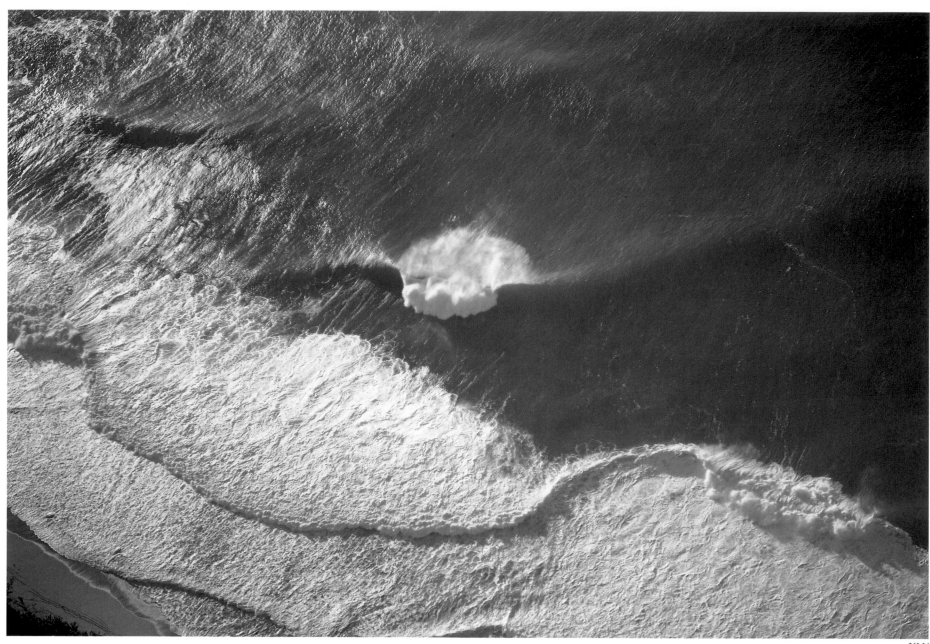

XLV

XLVI

XLVIII

IL

L

LI

The bleat, the bark, bellow, and roar
Are waves that beat on Heaven's shore.

—William Blake

LIII

LIV

LV

Sabrina fair,
   Listen where thou art sitting
   Under the glassy, cool, translucent wave…
          —John Milton

LVIII

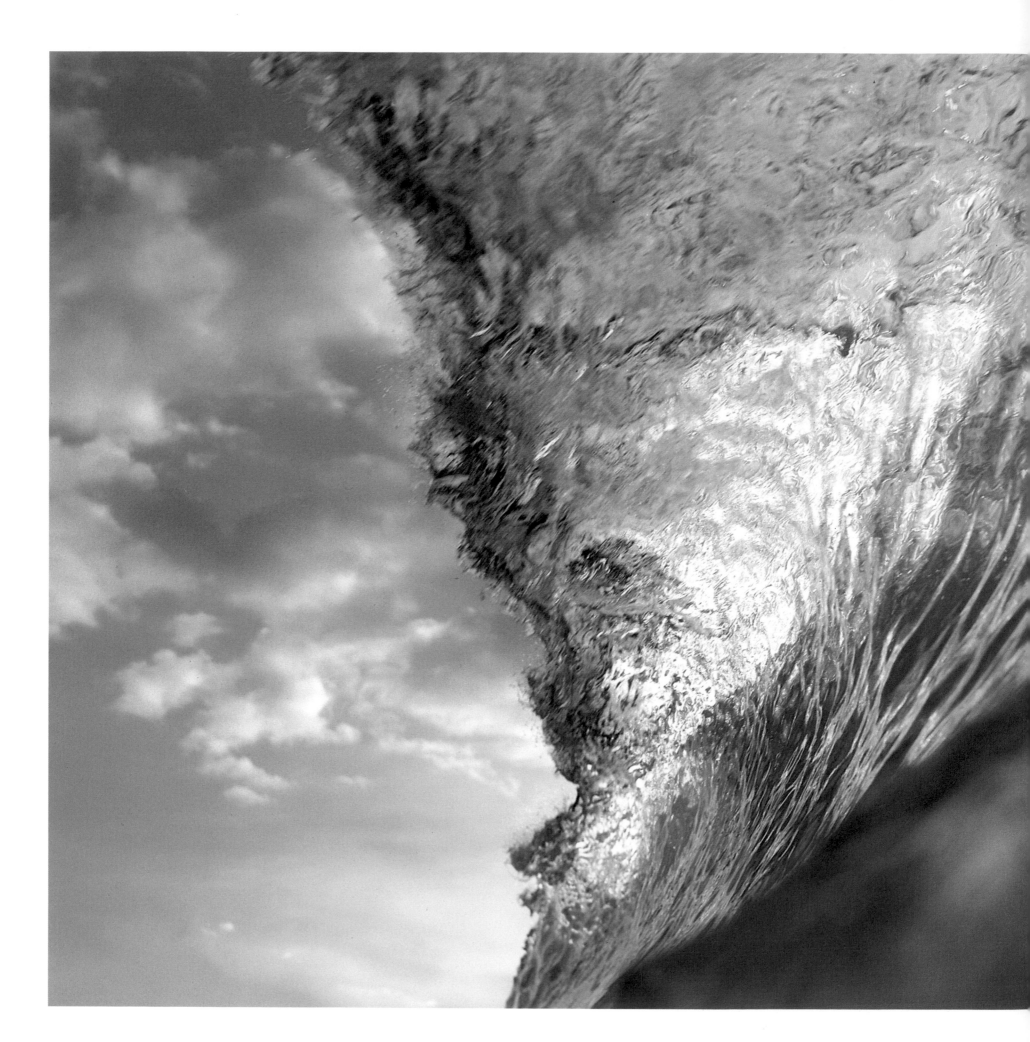

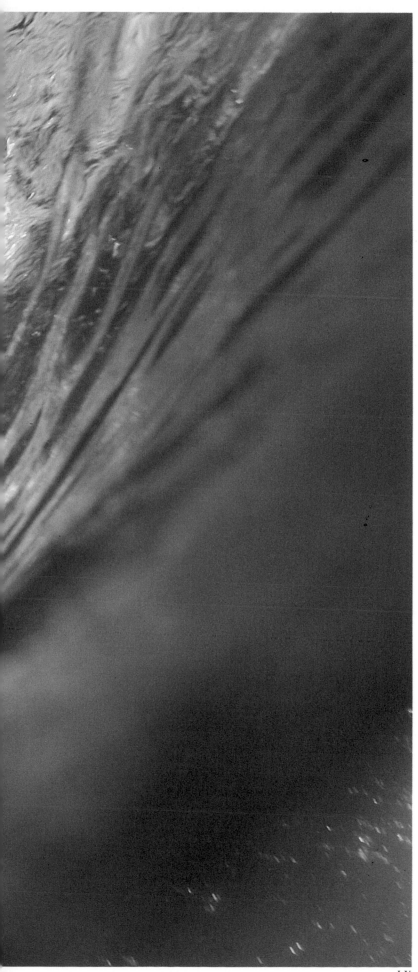

To me the sea is a continual miracle,
The fishes that swim—the rocks
*   —the motion of the waves*
*   —the ships with men in them,*
What stranger miracles are there?

—Walt Whitman

LXI

LXIII

LXIV

LXV

LXVI

LXVII

LXVIII

LXIX

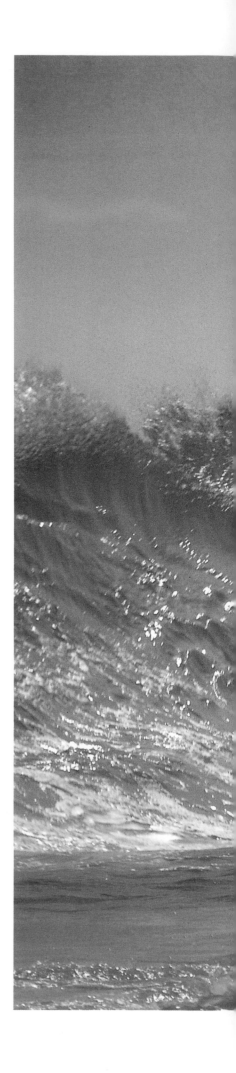

What are the wild waves saying
Sister, the whole day long,
That ever amid our playing,
I hear but their low lone song?

—Joseph Edwards Carpenter

LXXII

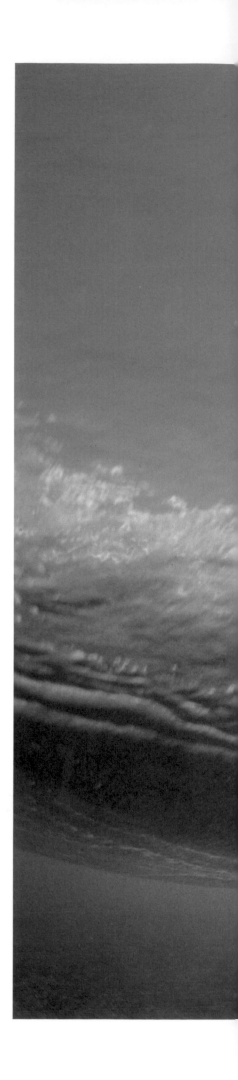

LXXIV

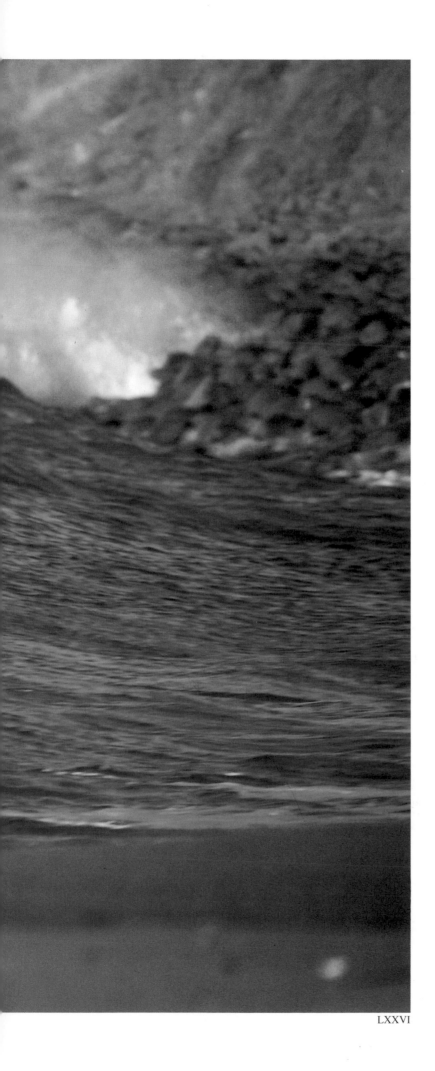

We are the bees of the invisible.
We distractedly plunder
the honey of the visible in order to
accumulate it within
the golden hive of the invisible.

—Rainer Maria Rilke

LXXVIII

LXXIX

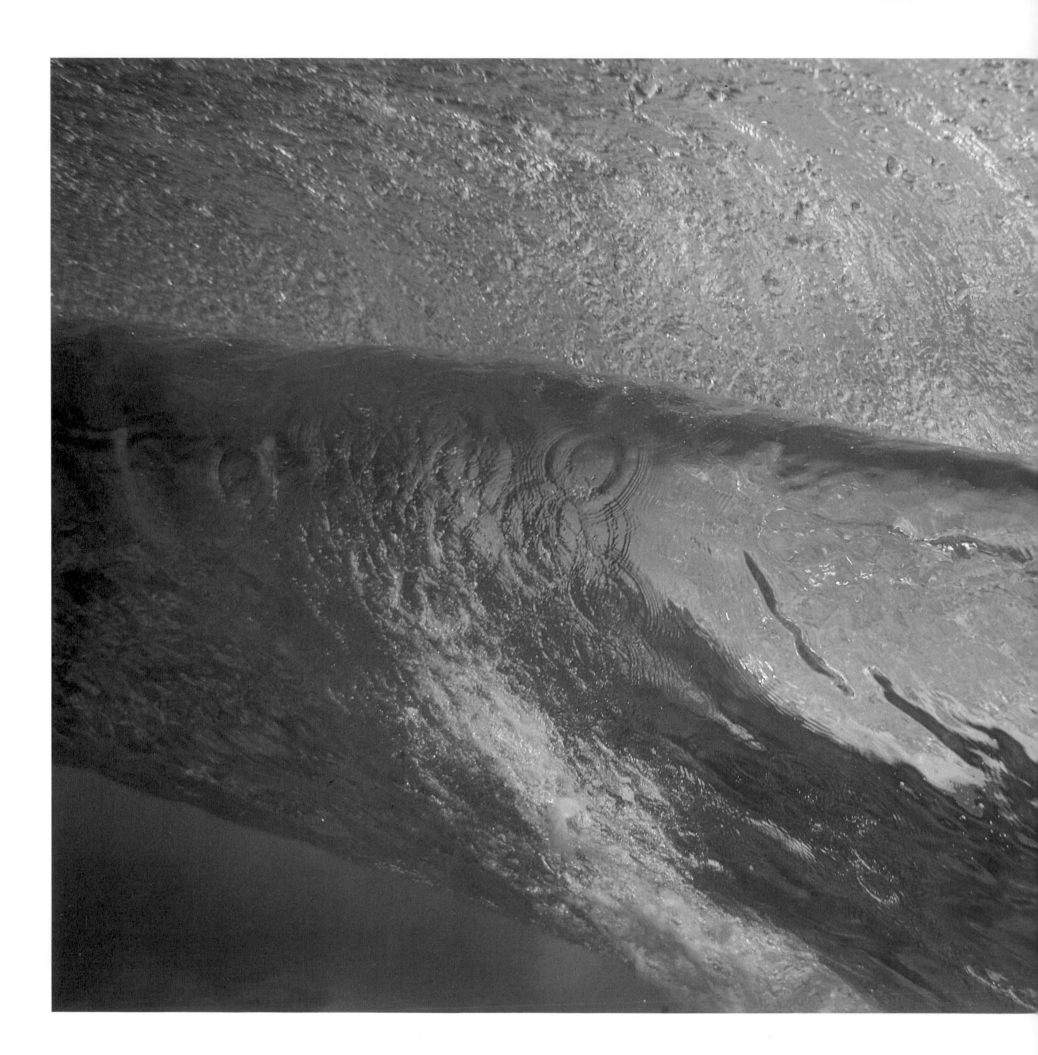

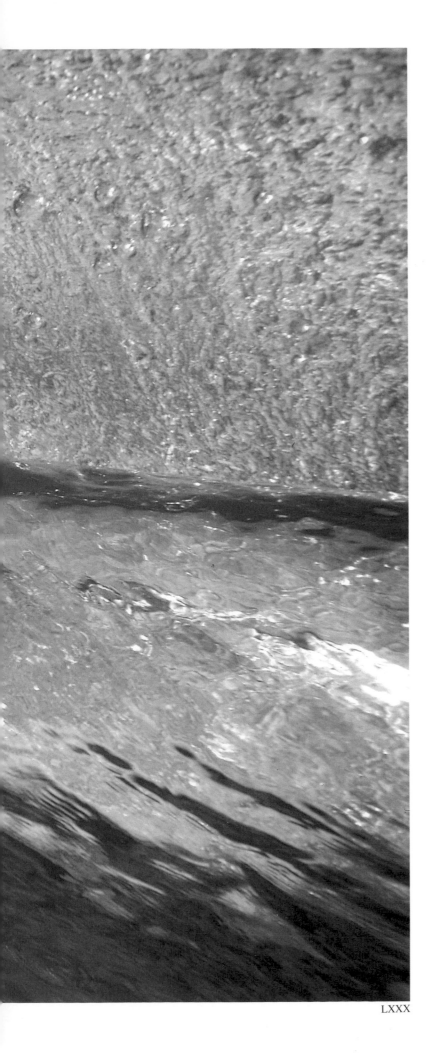

LXXX

LXXXI

LXXXII

LXXXIII

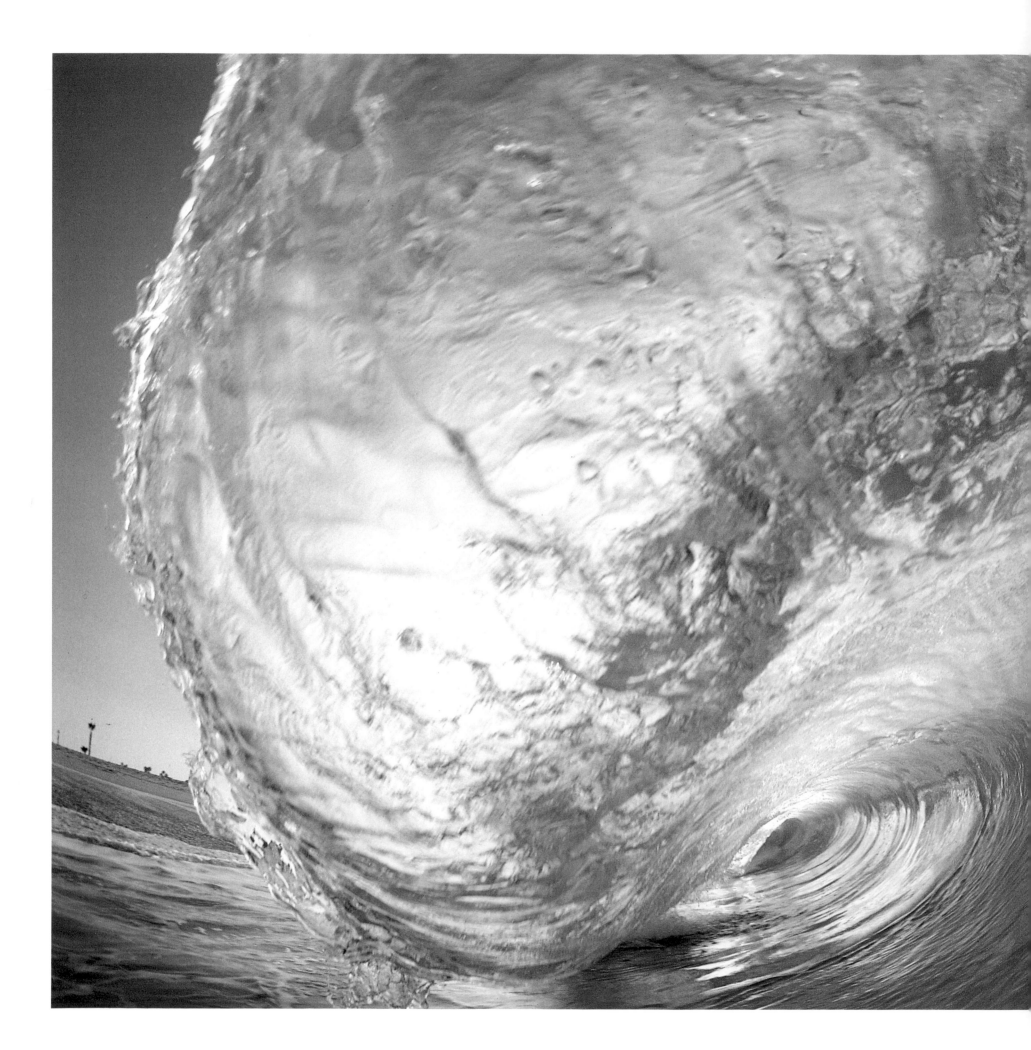

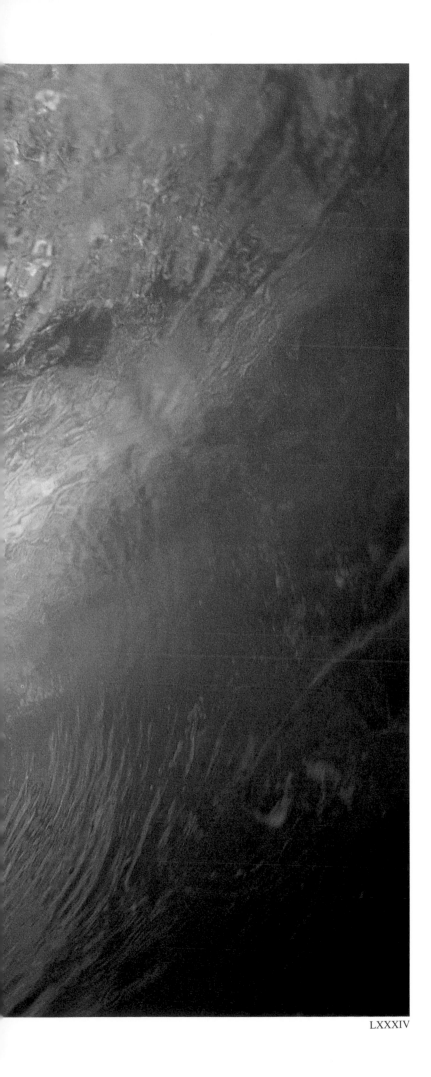

I see the waves upon the shore
Like light dissolved in star-showers, thrown.

—Percy Bysshe Shelley

LXXXV

LXXXVII

LXXXVIII

Now my brothers call from the bay;
Now the great winds shorewards blow;
Now the salt tides seawards flow;
Now the wild white horses play,
   Champ and chafe and toss in the spray.

—Matthew Arnold

XCI

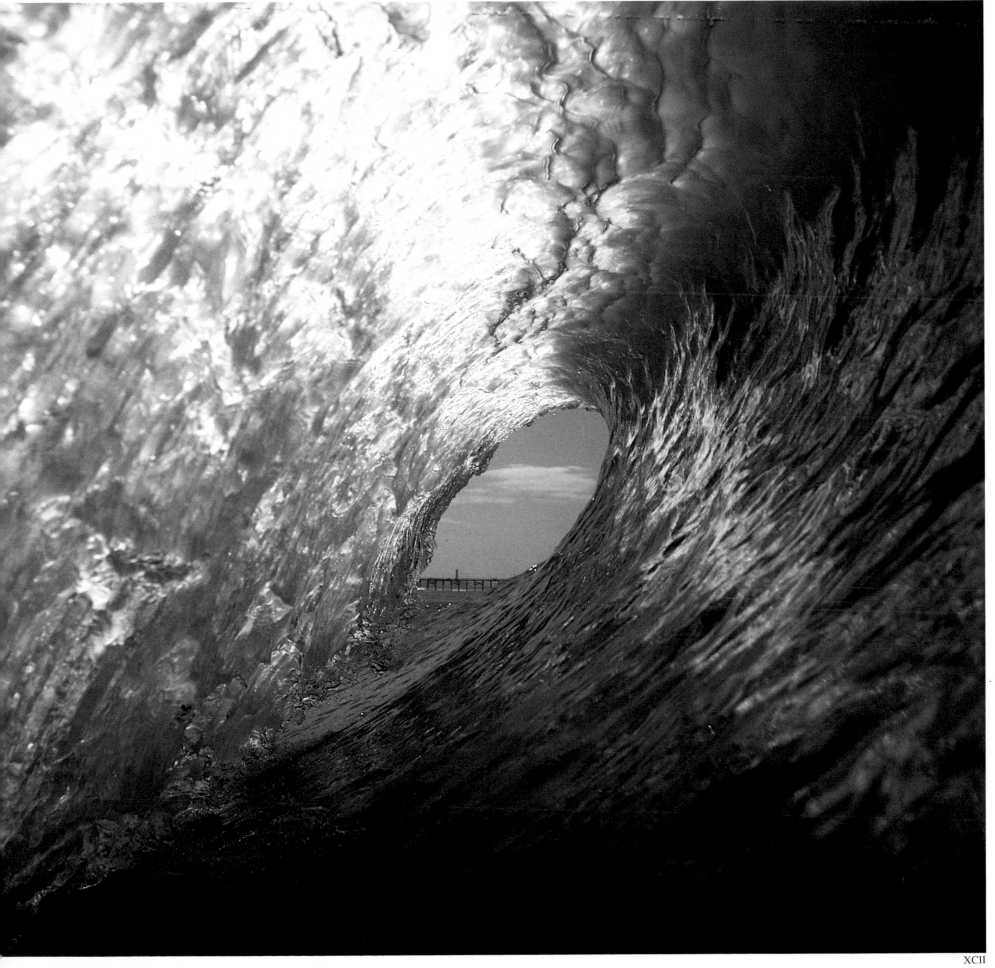

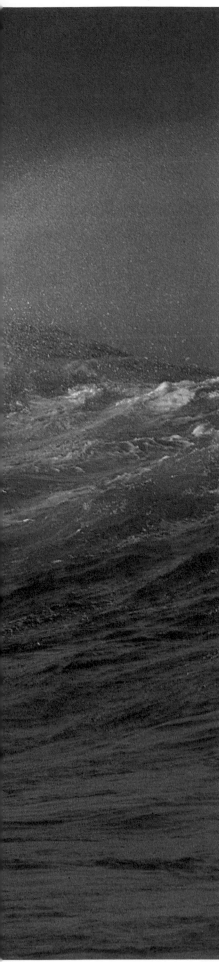

XCIII

XCIV

XCV

XCVI

XCVII

The sea never changes and its works,
for all the talk of men,
are wrapped in mystery.

—Joseph Conrad

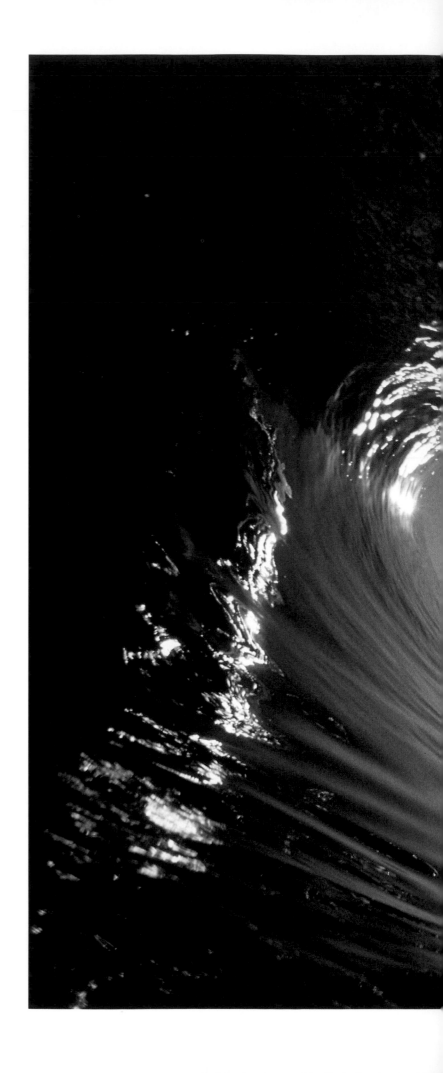

C

CI

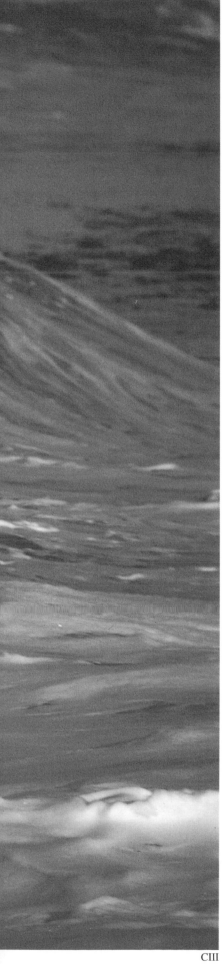

CIII

CIV

CV

The three great elemental sounds in nature
are the sound of the rain, the sound of the wind...
and the sound of outer ocean on a beach.
I have heard them all
and that of the ocean is the most
awesome, beautiful, and varied...

—Henry Beston

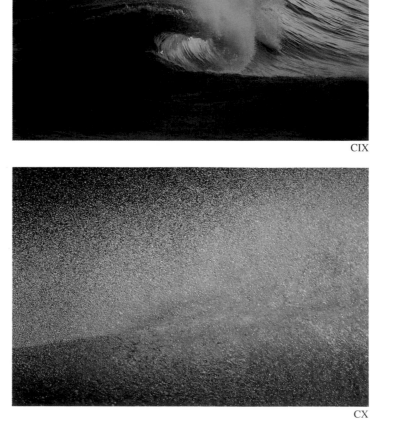

CVIII

CIX

CX

CXI

CXII

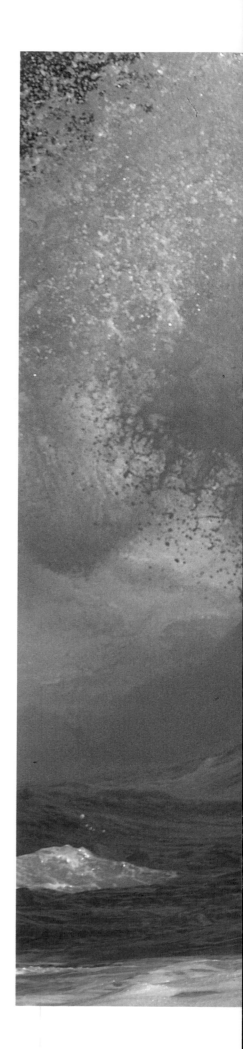

I have known the sea too long
    to believe in its respect for decency,

—Joseph Conrad

CXV

CXVI

CXVII

CXVIII

CXIX

Energy is Eternal Delight.

—William Blake

INDEX TO THE PLATES

I
North Shore Oahu, Hawaii
REGGIE HODGSON
Canon AE-1 • Century 650mm
Kodachrome 64 • f8 @ 1/500

"The shorebreak at Off-the-Wall near the Banzai Pipeline. In the spring when the sand moves back in with the north and northeast swells, the backwash gets so bad at times it's too weird to surf. But, then again, the crowds are long gone."

II
North Shore, Oahu, Hawaii
DENJIRO SATO
Nikonos • Nikkor 35mm
Kodachrome 64 • f8 @ 1/500

"Backdoor Pipeline."

III
Bayhead, New Jersey
DICK MESEROLL
Canon T-90 • 135mm w/85B warm filter
Kodachrome 64 • f11 @ 1/1000

"Early autumn swell lines and still warm water combine with a premature, cold-air front from Canada to form a 'steaming' ocean-mist effect as they interact in the dawn's light. There is little argument that this is 'wave' art."

IV
Waimea Bay, Oahu, Hawaii
DON KING
Canon F-1 • Canon 800mm
Kodachrome 64 • f6.3 @ 1/500

"This photo of the Waimea shorebreak really epitomizes the force and power of Waimea. It can be compared to the mightiest rapids in the Colorado River. I shot this from the safety of the shore."

V
North Shore Oahu, Hawaii
WARREN BOLSTER
Nikon F3 • Nikkor 135mm
Kodachrome 64 • f4.5 @ 1/1000

"On Jan. 5, 1985, while waiting for surfer Alec Cooke to catch a wave, we shot the outer reef swells to kill time. The waves were running in the 20- to 35-foot range (I know it doesn't look it!), and Cooke had to be airlifted out to the lineup to get through the close-out surf inside. It was fascinating to watch the waves bend as they refracted around the reefs. Outside Log Cabins, Third Reef Pipeline and Third Reef Rocky Point were all photographed on this day. Shooting from a helicopter can make waves look bigger or smaller depending on the angle/perspective. My interest as a photographer is more with the esoteric and ever-fascinating ocean angles, colors and moods. Size to me is not as important. This shot shows the swell adjusting to the reef at Outside Pipeline."

VI
Waimea Bay, Oahu, Hawaii
DON KING
Canon F-1 • Canon 800mm
Kodachrome 64 • f5.6 @ 1/500

"Just another afternoon at the Bay. Something like five broken boards today."

VII
The Azores
TONY ARRUZA
Nikon FM2 • Nikkor 180mm
Kodachrome 64 • f8 @ 1/500

"This view is from atop a 1,000' cliff at the west point of São Jorge Island."

VIII
Isla Natividad, Baja, Mexico
SONNY MILLER
Nikon N2000 • Nikkor 16mm fisheye
Kodachrome 64 • f4 @ 1/1000

"Pitch out! What I wanted out of this shot was the feeling you get when you're paddling out and a wave pitches out over you as you punch through."

IX
Los Angeles/Ventura County Line, California
WOODY WOODWORTH
Pentax Spotmatic • Pentax 300mm
Kodachrome X • f5.6 @ 1/500

"Getting an image like this is what photography of the ocean is all about."

X
Oahu, Hawaii
DENJIRO SATO
Nikon FE2 • Nikkor 135mm
Kodachrome 64 • f4 @ 1/500

"West side of Oahu."

XI
North Shore Oahu, Hawaii
DENJIRO SATO
Nikon FE2 • Nikkor 135mm
Kodachrome 64 • f5.6 @ 1/500

"The Banzai Pipeline."

XII
Ehukai Beach, Oahu, Hawaii
DON KING
Olympus • Zuiko 16mm
Kodachrome 64 • f5.6 @ 1/500

"Rolling thunder. You can feel the thunder rolling over you. The only way out is through the seams."

XIII
Newport Beach, California
DALE KOBETICH
Canon A-1 • Canon 15mm
Kodachrome 64 • f5.6 @ 1/500

"Shot at the famous Newport Wedge in water approximately six-inches deep while listening to Pink Floyd on my water-housed Walkman. Evening time. Fun!"

XIV
Earth
NASA
Hasselblad • 50mm
Ektachrome 100 (70mm) • (Unrecorded)

Shot by a member of the crew on Apollo 17, the last mission of NASA's moon program, four hours after launch and some 21,750 miles from Earth. Africa and much of Asia are visible. Photo was taken at 1:10 p.m. CST on December 7, 1972.

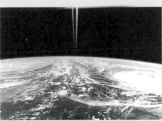

XV
Pacific Ocean
NASA
Nikkon • Nikkor 50mm
Ektachrome 100 (70mm) • (Unrecorded)

Two major tropical depressions over the Pacific-'on the left Typhoon Pat, on the right Typhoon Odessa. Taken in the first week of September 1985 from the Discovery, on the twentieth mission of the space shuttle.

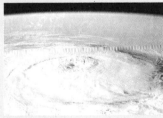

XVI
Caribbean Sea
NASA
Hasselblad • 50mm
Ektachrome 100 (70mm) • (Unrecorded)

Also from mission twenty of the space shuttle, taken from Discovery, Hurricane Elena spirals below on September 2, 1985.

XVII
The Gulf of Mexico
NASA
Hasselblad • 50mm
Ektachrome 100 (70mm) • (Unrecorded)

Taken on the last successful voyage of the Challenger during the week of October 30-November 6, 1985. Hurricane Juan had been downgraded to a tropical storm. Taken just south of Mobile, Alabama.

XVIII
Waimea Bay, Oahu, Hawaii
WARREN BOLSTER
Nikon F3 • Century 650mm
Kodachrome 64 • f6.3 @ 1/500

"Waimea shorebreak double-up power dump. A study in ocean power at a respecful distance. Never-ending study in power and wave form with a dark background in early morning light. Certain death for the unwary bodysurfer or surfer. Getting through the shorebreak on a 20´ to 25´ day like this can be the timing call of a lifetime. Fortunately it's a lot safer closer to the road where this was taken."

XIX
Barbados
DAVID DiGIROLAMO
Canon A-1 • Canon 70-210mm
Kodachrome 64 • f11 @ 1/250

"Native surfer."

XX
Newport Beach, California
MIKE MOIR
Pentax ME • Pentax 20mm
Kodachrome 64 • f5.6 @ 1/500

"Brian 'Tank' Dawson (or his brother) at the Newport Wedge in the early '80s. Perhaps the purest way to ride a wave is bodysurfing."

XXI
Sandy Beach, Oahu, Hawaii
WARREN BOLSTER
Nikon F3 • 16mm Nikkor
Fuji 100D • f5.6 @ 1/1000

"I've still got a bad back from shooting here. For a shorebreak wave, this is one of the most difficult to shoot. The secret is to let the current pull you out in a prone position. You have to stay very, very low or the wave will dump you. You've got to time it perfectly to be sucked out, but not over. A two-foot wave here literally almost broke my back." Bodyboarder in a tight spot.

XXII
Waimea Bay, Oahu, Hawaii
DENJIRO SATO
Nikon FE2 • Nikkor 135mm
Kodachrome 64 • f8 @ 1/500

"Grace under pressure."

XXIII
Makaha, Oahu, Hawaii
MIKE MOIR
Pentax Spotmatic • Vivitar 400mm
Kodachrome II • f6.3 @ 1/500

"Makaha short stack. Drubbing in aqua-blue."

XXIV, XXV, XXVI
Waimea Bay, Oahu, Hawaii
GORDINHO
Nikon FM • Century 1000mm
Kodachrome 64 • f8 @ 1/500

"Every picture tells a story. This one tells the story of a bad moment on a big day at Waimea Bay. I shot this from the rock on the west side of the Bay. This perspective gave me a low angle that looks almost like the shot was taken from the water."

XXVII
North Shore Oahu, Hawaii
GORDINHO
Nikon FM • Century 1000mm
Kodachrome 64 • f8 @ 1/500

"It was only some 30 years ago that surfers first attempted to ride the Banzai Pipeline. Up to then, the word was that to surf there was to die. The wave is powerful, savagely hollow, and the bottom is a minefield of coral heads. To see Robby Naish, the world's best board-sailor, sail out and shred the place was one of the greatest things I've ever seen."

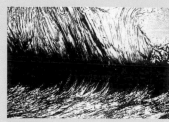

XXVIII
Newport Beach, California
DENNIS JUNOR
Canon A-1 • Canon 300mm
Kodachrome 64 • f8 @ 1/500

"Golden lip splash!"

XXIX
Waimea Bay, Oahu, Hawaii
DENJIRO SATO
Nikon FE2 • Nikkor 600mm
Kodachrome 64 • f8 @ 1/500

"Study in power and finesse."

XXX
Newport Beach, California
WOODY WOODWORTH
Nikon FM • Century 650mm
Kodachrome 64 • f22 @ 1/30

"The Wedge. You pre-focus, then use a cable release."

XXX
San Clemente, California
GUY MOTIL
Contax RTS II • Zeiss 35mm
Kodachrome 64 • f16 @ 1/4

"Winter's lowest tides, around sunset. I'm wet, cold and happy. The pleasure seems to balance out the discomfort."

XXXII
San Clemente, California
DENNIS JUNOR
Canon A-1 • Canon 50mm
Kodachrome 64 • f11 @ 1/500

"Aerial view of massive swell lines rolling in? About 10 to 12 feet if you were an ant. Interesting pattern, one that any surfer can relate to when it's dead flat and conditions are perfect and you wish you were an inch tall."

XXXIII
San Clemente, California
GUY MOTIL
Contax RTS II • Zeiss 300mm
Kodachrome 64 • f8 @ 1/8

"Taken at sunset, the sun reflecting on the back of the wave. Almost like wet fire. Visual signals sometimes fool the mind's eye."

XXXIV
North Shore Oahu, Hawaii
DENJIRO SATO
Nikon FE2 • Nikkor 24mm
Kodachrome 64 • f5.6 @ 1/500

"Backdoor Pipeline."

XXXV
Newport Beach, California
MIKE MOIR
Canon T-70 • Canon 15mm
Kodachrome 64 • f5.6 @ 1/500

"Moment of backwash impact, circa 1983."

XXXVI
Jalama Beach, California
CRAIG PETERSON
Nikon F • Nikkor 300mm
Kodachrome 64 • f4.5 @ 1/250

"Dividing lines between tides."

XXXVII
North Shore Oahu, Hawaii
DENJIRO SATO
Nikon FE2 • Nikkor 600mm ED
Fujichrome 50D • f11 @ 1/500

"North Shore."

XXXVIII
Puerto Rico
DICK MESEROLL
Nikon FM • Nikkor 85mm
Kodachrome 64 • f3.5 @ 1/500

"A spot called Gas Chambers. Pound for pound, pounding for pounding, one of the unrecognized heavyweight waves in the world. Few are called, fewer are ridden. A beautiful, yet hideous sight to behold when the backwash is working."

XXXIX
Tahiti
BERNIE BAKER
Nikon FM • Century 1000mm
Kodachrome 64 • f8 @ 1/500

"I shot two rolls of film with this composition and each one was identical. A dream wave to surf and impossible to make."

XL
Fiji
CRAIG PETERSON
Nikon F2 • Nikkor 105mm
Kodachrome 64 • f5.6 @ 1/500

"South Pacific swell showers—a study in blue at Cloudbreak Reef, Fiji."

XLI
North Shore, Oahu, Hawaii
DENJIRO SATO
Nikon FE2 • Nikkor 300mm f4.5 ED
Kodachrome 64 • f5.6 @ 1/500

"Banzai Pipeline."

XLII
Newport Beach, California
WOODY WOODWORTH
Nikon FM • Nikkor 20mm
Kodachrome 64 • f5.6 @ 1/500

"You take a beating doing this type of angle at the Newport Wedge."

XLIII
San Clemente, California
GUY MOTIL
Contax TRS II • Zeiss 85mm
Kodachrome 64 • f11 @ 1/30

"From my 'Wet Light Series.' Motion subtracts detail, allowing direction of water movement to be seen more clearly. Light (in the impressionist tradition) is also more clearly seen without detail. Water is sometimes opaque, transparent, reflective, or a combination of all of these. This is especially noticeable when a wave is breaking—this is my concept of 'wet light'."

XLIV
Fernando de Noronha, Brazil
ROBERT BECK
Canon T-90 • Canon 15mm
Kodachrome 64 • f5.6 @ 1/500

"Fierce beachbreak on tiny island off northern coast of Brazil. Shallow sandy bottoms make every wave hollow. Minimum surfers...maximum tube time."

XLV
North Shore Oahu, Hawaii
WARREN BOLSTER

Pentax MX • Takumar 135mm
Kodachrome 64 • f4.5 @ 1/1000

"One of my first aerial photo sessions. Pipeline was so big and shallow this day, only two surfers went out and promptly came in, although it looked perfect. A good day for the helicopter perspective. This was taken from about 1,500 feet, and the surf was about 12 feet. We shot high and low angles, and later a couple of surfers ventured back out for some low-level shots. The pilot is no longer living, and it's a tribute to Lt. Col. Powell Moore that he had the vision to consistently put me in positions like this."

XLVI
North Shore Oahu, Hawaii
JON FOSTER

Nikon FE2 • Nikkor 28mm-80mm zoom
Kodachrome 64 • (Unrecorded)

"Surge and flow."

XLVII
San Clemente, California
GUY MOTIL

Contax RTS II • Zeiss 35mm
Kodachrome 64 • f5.6 @ 1/60 w/flash

"Taken on the day of the winter solstice, using a flash and the ambient light from the twilight sky a little after sunset. Combination of motion and frozen action. Reminds me of piercing ice and bitter cold, though this evening was really quite pleasant. From the 'Wet Light Series.'"

XLVIII
South Coast, England
NOEL ROSSI

(Technical information unrecorded)
(See following description)

XLIX
South Coast, England
NOEL ROSSI

(Technical information unrecorded)

"These shots were taken in very rare 'perfect' conditions of low sun, wind, a nice sky and rough seas. Used a red filter to cut down the blue. Lots of spray in the wind, so bad to constantly check for clear lens."

L
Tortola, British Virgin Islands
WOODY WOODWORTH

Nikon FE2 • Nikkor 20mm
Kodachrome 64 • f5.6 @ 1/500

"Same session as the other wave shot." (See Plate LXXXIX)

LI
North Shore Oahu, Hawaii
DENJIRO SATO

Nikon FE2 • Nikkor 24mm
Kodachrome 64 • f5.6 @ 1/500

"Backdoor Pipeline."

LII
Oahu, Hawaii
DENJIRO SATO

Nikon FE2 • Nikkor 300mm ED
Kodachrome 64 • f5.6 @ 1/500

"West side Oahu."

LIII
Zuma Beach, California
STEVE SAKAMOTO

Canon F1 • Canon 600mm
Kodachrome 64 • f22 @ 1/15

Late afternoon and slow shutter speed.

LIV
San Clemente, California
GUY MOTIL

Contax RTS II • Zeiss 300mm f4
Kodachrome 64 • f16 @ 1/60

"Early morning sunlight (8:00 a.m.) on wave face. Taken in same location and manner as LXV, but on a clear day. From the 'Wet Light Series.'"

LV
Huntington Beach, California
WOODY WOODWORTH

Nikon FE2 • Century 650mm
Kodachrome 64 • f6.3 @ 1/1000

"From the Huntington Beach Pier looking south. Shot the moment the sun broke over the horizon."

LVI
Hermosa Beach, California
LEROY GRANNIS

Pentax ME • Pentax 28mm
Kodachrome 64 • f58 @ 1/250

"The south side of the Hermosa Beach Pier. About 6-8′ perfect shape. Great reflection. Probably too fast to ride."

LVII
Laguna Beach, California
WOODY WOODWORTH

Pentax Spotmatic • Vivatar 20mm
Kodachrome 64 • f5.6 @ 1/1000

"One of my first really nice wave shots."

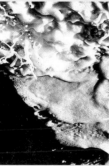

LVIII
Laguna Beach, California
ART BREWER

Nikon F3T • Nikkor 300mm EDIF
Kodachrome 64 • f4.5 @ 1/500

"Frozen-motion whitewater by the Aliso Beach Pier."

LIX
Cape Regina, New Zealand
PETER SIMONS

Nikon FM2 • Nikkor 105mm
Kodachrome 64 • f5.6 @ 1/250

"Parallel with Hawaiian lore of Oahu's Kàena Point, Cape Regina is the departure place for Maori souls leaving this earthly realm. The waves here remind me of drifting, searching souls, although I feel I could have shot a better photo."

LX
North Shore Oahu, Hawaii
GRAIG FINEMAN

Nikon FE2 • Nikkor 35mm
Kodachrome 64 • f5.6 @ 1/500

"The Banzai Pipeline."

LXI
North Shore Oahu, Hawaii
CRAIG FINEMAN

Nikon F3 • Nikkor 600mm f5.6 EDIF
Kodachrome 64 • f6.3 @ 1/500

"Precision Pipeline."

LXII
North Shore Oahu, Hawaii
WARREN BOLSTER

Pentax MX • Takumar 50mm macro
Kodachrome 64 • f6.3 @ 1/500

"Power sweep at the Banzai Pipeline. Photo taken during the Pipeline Masters surfing competition of one of the few waves that nobody caught during finals. Pipeline was classic this day. I remember being timid because I was using such a short lens in competition in eight-to-ten-foot surf."

LXIII
North Shore Oahu, Hawaii
BUD McCRAY

Canon T-90 • Canon 800mm
Kodachrome 64 • f5.6 @ 1/350

"This shot of a 25-foot wave at Kammieland was taken from the shore at Sunset Beach. The swell was 18 to 25 feet with perfect conditions. West swells would peel from outside Rocky Point with Kammie's as the bowl. Some would finish off by closing out across Sunset. The soft glow of the diminishing light made it look rideable, but the reality was that any attempt could be fatal. This wave ground off, then closed outside of Sunset. The soup washed up over the beach and onto the highway."

LXIV
San Clemente, California
GUY MOTIL

Contax RTS II • Zeiss 35mm f1.4
Kodachrome 64 • f8 @ ~1/2

"From my 'Wet Light Series.' Complementary color scheme (blue vs. pink)…tranquility, serenity, day's end. I'm ready to head home."

LXV
San Clemente, California
GUY MOTIL

Contax RTS II • Zeiss 300mm
Kodachrome 64 • f22 @ 1/30

"Sunlight filtered through the smoke from a fall brush fire. The wave brings to my mind a 'parchment palimpsest'—wave on wave, surfer's track on surfer's track, photo on photo. Taken at virtually the same spot and manner as LIV."

LXVI
Carlsbad, California
ROBERT BECK

Canon F-1 • Canon 300mm
Kodachrome 64 • f5.6 @ 1/500

"Ponto. My home spot. I check it almost every day. Dolphins bodysurfed from south end of the beach to the north. I followed them along the shore. These two tandemed the wave of the day."

LXVII
San Onofre, California
JEFF DIVINE

Nikon FE2 • Nikkor 135mm
Kodachrome 64 • f6.3 @ 1/500

"The breakfast patrol at Lower Trestles. Everyone who surfs has seen the pelicans wing-tipping across the wave faces on fish reconnaissance. I just happened to have a camera in my hand as they swooped by."

LXVIII
Newport Beach, California
MIKE MOIR

Canon T-90 • Canon 300mm L Series
Kodachrome 64 • f2.8 @ 1/1000

"You could visualize stuffing the right pocket, but look at the twist in the section ahead of you. Fluorite glass yields pin-sharp tele results. The Newport Peninsula."

LXIX
Newport Beach, California
MIKE MOIR

Canon T-90 • Canon 300mm L Series
Kodachrome 64 • f2.8 @ 1/1000

"On the Newport Peninsula. Chiropractor's delight, but the vacuum-pack pulling in is incredible."

LXX
Puerto Azul, Mexico
CRAIG PETERSON

Nikon F • Century 650mm
Kodachrome 64 • f8 @ 1/500

"Cannibal shorebreak caught in a feeding frenzy. Southern Mexico."

LXXI
Shark Island, Australia
PETER SIMONS

Nikonos • 35mm UW
Kodachrome 64 • f5.6 @ 1/125

"The sheer mysticism of a breaking wave—the slower shutter speed conveys this feeling."

LXXII
Durban, South Africa
TONY ARRUZA

Nikon FE • Nikkor 300mm f/4.5
Kodachrome 64 • (Unrecorded)

"Sunrise lines. A big swell coming into Bay of Plenty at sunrise. Shot from the 17th floor of the Elangeni Hotel."

LXXIII
North Shore Oahu, Hawaii
DON KING

Olympus • 16mm
Kodachrome 64 • f5.6 @ 1/500

"Through the looking glass. The cylindrical motion of a hollow wave is even more apparent underwater. From this angle, it comes to life, spinning as the sunlight dances after it."

LXXIV
Newport Beach, California
WOODY WOODWORTH

Nikon FM • Nikkor 20mm
Kodachrome 64 • f5.6 @ 1/1000

"I've always loved shooting from the air. This was from 2,500 feet—a hurricane swell in October 1977."

LXXV
Laguna Beach, California
WOODY WOODWORTH

Pentax Spotmatic • Vivitar 20mm
Kodachrome 64 • f5.6 @ 1/1000

"Every winter (October to February) the sun rises in the eye of the wave."

LXXVI

Los Angeles/Ventura County Line, California
STEVE SAKAMOTO

Canon F-1 • Canon 800mm
Kodachrome 64 • f4 @ 1/250

"Summer surge and California tones."

LXXVII

Bells Beach, Victoria, Australia
STEPHEN RYAN

Nikon F3 • Nikkor 50mm
Kodachrome 64 • f4 @ 1/500

"This Bells Beach storm overview was taken on December 7, 1986. It was the day after my birthday, and the air temperature was 32°C (90°F). I was standing there and it was like God put a spotlight on six-foot Bells. (I was actually trying to get eight-foot Bells, but I didn't have any film in the camera until it was six-foot.)"

LXXVIII

North Shore Oahu, Hawaii
MIKE WAGGONER

Nikon FM2 • Nikkor 105mm
Kodachrome 64 • f5.6 @ 1/500

"Banzai Pipeline, shot in February or March 1988. The beauty at Pipe is awesome! On this day the waves were pitching out rather far. Vince Cavataio, Bernie Baker and myself were having a great time shooting that day. Mahalo to all the water photographers who have gone before me and who have inspired my work. Mahalo also to our mother ocean, who let me share that day and also let me capture it on film."

LXXIX

North Shore Oahu, Hawaii
GUY MOTIL

Contax RTS II • Zeiss 300mm x 1.4 tele.
Kodachrome 64 • f5.6 @ 1/500

"Pipeline, with big surf and strong winds near sunset."

LXXX

North Shore Oahu, Hawaii
WARREN BOLSTER

Pentax MX • Takumar 17mm
Kodachrome 64 • f4.5 @ 1/1000

"My guess is that this underwater tube is Backdoor Pipeline or Off-the-Wall with a rare sandbar breaking on a north swell. Esoterically, I like underwater shots as much as anything. When the bottom is sand and the day is clear, faster shutter speeds reveal more than your eye can see. Getting the photos back, especially the solo tubes, is always entrancing."

LXXXI

North Shore Oahu, Hawaii
ROBERT BECK

Canon F-1 • Canon 300mm
Kodachrome 64 • f4 @ 1/500

"Late afternoon session at Pipeline."

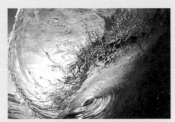

LXXXII

North Shore Oahu, Hawaii
DON KING

Olympus • Olympus 300mm
Kodachrome 64 • f4.5 @ 1/500

"The limelight. I was standing waist-deep on a shifting sandbar, waiting for the light to get perfect, and the tide kept coming up. Right at sunset the tide was up to my neck (I didn't have a waterproof housing), and my wife swam my camera bag out to me so I could change lenses and get the shot. As it got dark, I put the camera bag on a friend's surfboard and prayed I'd get across to the beach before I lost it."

LXXXIII

Oxnard, California
ROB KEITH

Nikon FM2 • Nikkor 135mm
Kodachrome 64 • f4.5 @ 1/500

"I was in a small single-engine Cesna at 100 feet on a routine wave-observation flight based out of San Diego. It was generally flat from San Diego to the northern L.A. County line that morning…until we passed Point Magu, and there it was—the beginning of a new northwest power swell just hitting the coast at Oxnard Shores. It was seven to ten feet for the next three days."

LXXXIV

Newport Beach, California
MIKE MOIR

Canon T-90 • Canon 15mm
Kodachrome 64 • f4 @ 1/500

"Newport Peninsula, 1987. A sunny dawn is the ideal time for wave sculptures."

LXXXV

Santa Cruz, California
DENNIS JUNOR

Canon A-1 • Canon 50mm
Kodachrome 64 • f5.6 @ 1/500

"Sea fury. Evokes the feeling of the 'Wild North Coast.' Massive surf rolling over anything in its path. A true display of the ocean's power."

LXXXVI

Oxnard, California
CRAIG FINEMAN

Nikon FTN • Century 650mm
Kodachrome 64 • f6.3 @ 1/500

"Wreck of La Janelle *off Oxnard."*

LXXXVII

Waimea Bay, Oahu, Hawaii
CRAIG FINEMAN

Nikon F3 • Nikkor 600mm EDIF
Kodachrome 64 • f6.3 @ 1/500

"Pacific Power."

LXXXVIII

Santa Cruz, California
WOODY WOODWORTH

Nikon FE2 • Nikkor 300mm
Kodachrome 64 • f5.6 @ 1/500

"Taken between Mitchell's Cove and Steamer Lane; this area is incredible for crashing waves and dramatic light."

LXXXIX

Tortola, British Virgin Islands
WOODY WOODWORTH

Nikon FE2 • Nikkor 20mm
Kodachrome 64 • f5.6 @ 1/500

"When I went to the Carribean I wanted to make one photo that really captured the clarity and color of the water. I was 100% satisfied with this shot."

XC

Waimea Bay, Oahu, Hawaii
WARREN BOLSTER

Nikon F3 • Century 650mm
Kodachrome 64 Pro • f6.8 @ 1/500

"Waimea shorebreak explosion—triple-upped backwash. Early morning Waimea shorebreak offers a unique opportunity to use a long telephoto on a front-lit tube with a side-view angle against a dark black background. The variety of wave action here is perhaps the most complicated and inspiring power-wise in the world. I consider this one of my best four phots. If I were going to select one photo to represent my work as a poster, this would be the one."

XCI
Hampton Beach, New Hampshire
MICHAEL BAYTOFF
Nikon F3T • Nikkor 300mm
Fujichrome 50D • f8 @ 1/250

"Cormorants on the wing in late fall."

XCII
Oceanside, California
GUY MOTIL
Contax RTS II • Zeiss 35mm f1.4
Kodachrome 64 • f5.6 @ 1/500

"A frozen moment. The photo allows you to scrutinize what in real life moves too quickly to see in any detail."

XCIII
Waimea Bay, Oahu, Hawaii
DON KING
Olympus • Olympus 300mm
Kodachrome 64 • f4.5 @ 1/500

"TNT! A true gut-wrenching barrel. The total chaotic power gives you the feeling you're seeing the aftermath of Grand Coulee Dam breaking."

XCIV
Laguna Beach, California
WOODY WOODWORTH
Nikon FE2 • Century 650mm
Kodachrome 64 • f6.3 @ 1/1000

"There is a high cliff in the background that stays in the shadow as the mid-morning sun catches the wave spray."

XCV
Zuma Beach, California
STEVE SAKAMOTO
Canon F-1 • Canon 400mm
Kodachrome 64 • f8 @ 1/500

"Early morning peaks."

XCVI
Cayucos, California
MIKE MOIR
Pentax ME • Vivitar 20mm
Kodachrome 64 • f8 @ 1/500

"Bone-numbing cold water (low 50s, maybe high 40s). The sun even appears more distant with crisp four-foot waves and offshore winds all day. This was during a surf session with Dave Parmenter and friends at Cayucos Pier in November '81."

XCVII
Laguna Beach, California
WOODY WOODWORTH
Pentax Spotmatic • Vivitar 20mm
Kodachrome 25 • f5.6 @ 1/500

"Shot at the El Moro Trailer Park. This was a very strong Santana wind in December. This wave slammed me off the bottom harder than any other wave. The tide was a very negative low."

XCVIII
Santa Cruz, California
WOODY WOODWORTH
Nikon FE2 • Nikkor 20mm
Kodachrome 64 • f5.6 @ 1/1000

"Pleasure Point in Santa Cruz. I had shot many pictures with this angle 10 years earlier in Laguna Beach. I had just finished making a new water-housing and had a vision of this shot the night before. Got up the next morning and tested the housing."

XCIX
North Shore Oahu, Hawaii
WARREN BOLSTER
Nikon N2000 • Nikkor 20mm
Kodachrome 64 • f4.5 @ 1/2000

"Splashdance on the North Shore at Pipeline. First swell of spring and the swell had cut a ridge in shoreline sand. Pipeline was five to six feet all day. Kona variable clouds were moving in fast so I shot entire roll each time the sun came out. Seeing the beautiful day quickly turning dark with the approaching clouds, I tried to combine the foreground and background surf."

C
Baja
WOODY WOODWORTH
Pentax Spotmatic • Takumar 135mm
Kodachrome 64 • f5.6 @ 1/500

"This wave is six inches high. There were so many of these little beauties, all perfect! But I've never seen a beach do this again so perfectly."

CI
Fiji
DON KING
Olympus • Zuiko 16mm
Kodachrome 64 • f5.6 @ 1/500

"Taken at the surf spot called Cloudbreak, this view of a perfect wave from underwater resembles clouds forming in time-lapse of fast-motion photography. From this angle, a surfer riding by on the wave really looks like he's riding on clouds."

CII
Encinitas, California
WARREN BOLSTER
Pentax MX • Takumar 17mm
Kodachrome 64 • f4.5 @ 1/1000

"The beauty and power of a wave without any surfers around is one of the things I enjoy most."

CIII
La Jolla, California
RICK DOYLE
Canon F-1 • Canon 800mm
Kodachrome 64 • f5.6 @ 1/500

"Taken at Windansea Beach, the wave face is 12 feet! This photo's texture has an almost airbrushed feel to it. The high-tide backwash into the shore-break made the surf jack up, causing this condition."

CIV
North Shore Oahu, Hawaii
JEFF DIVINE
Nikon FE2 • Nikkor 24mm
Kodachrome 64 • f5.6 @ 1/30

"Log Cabins, near Pipeline."

CV
Waimea Bay, Oahu, Hawaii
GRAIG FINEMAN
Nikon F3 • Nikkor 600mm EDIF
Kodachrome 64 • f8 @ 1/500

"A swirling sweep of sea."

CVI
Moss Landing, California
WOODY WOODWORTH
Nikon FM • Century 650mm
Kodachrome 64 • f6.3 @ 1/500

"I was so excited shooting photos that I left my surfboard by the side of the car and drove off without it. This is a very, very cold Monterey Bay wave."

CVII
North Shore Oahu, Hawaii
STEVE WILKINGS
Pentax Spotmatic • Pentax 135mm
Kodachrome 25 • f5.6 @ 1/500

"This was one of those 'wow' photos. I knew it would be a 'nice' photo. It was my first Surfer Magazine poster and one of my all-time favorites."

CVIII
Huntington Beach, California
ROB GILLEY
Nikon N2000 • Nikkor 600mm
Kodachrome 64 • f5.6 @ 1/250

"Birth. Photo taken from the Huntington Pier at dawn in February. Two-to-four-foot swell. Very strong Santana condition."

CIX
Newport Beach, California
ROB GILLEY
Nikon FM2 • Nikkor 200mm
Kodachrome 64 • f4 @ 1/250

"A Wedge thickie. Photo taken from the jetty at sunset. A six-to-seven-foot south swell. August."

CX
Huntington Beach, California
MIKE MOIR
Pentax Spotmatic • Vivitar 400mm (+2x)
Kodachrome 64 • f13 @ 1/500

"Effervescent! I like a wave photo that looks like it's something else, or at least sparks the mind to a different direction. Like this one. A fiery furnace? Fizz from a freshly poured soft drink? The south side of Huntington Pier, about 1978."

CXI
North Shore Oahu, Hawaii
WARREN BOLSTER
Pentax MX • Takumar 50mm
Kodachrome 64 • f4.5 @ 1/1000

"A low-level aerial of Pipeline on a deserted, dangerous day. I remember a lot of activity in the helicopter while flying along the wave ridge. We caught both spray and updraft. This is a massive wave beginning to hit bottom and suck hard over very shallow reef—a ten-to-twelve-foot day. I was struggling to keep the lens clean of spray and my equipment from flying out of the helicopter, while the pilot struggled for control in the tradewinds to hold the angle."

CXII
Oahu, Hawaii
PETER CRAWFORD
Nikonos • Nikkor 35mm
Kodachrome 64 • f5.6 @ 1/500

"This was taken near a spot called Third Dip out near Yokohama on the West Shore past Makaha. I got some pretty good shots here, although it's not a very surfable wave as such. It's too whompy. I got some good barrel shots of some surfers, but this one happened to have nobody on it."

CXIII
Makaha, Oahu, Hawaii
JEFF DIVINE
Nikon FE2 • Nikkor 300mm
Kodachrome 64 • f5.6 @ 1/500

"The Makaha shorebreak."

CXIV
Shark Island, Australia
PETER SIMONS
Nikon FM2 • Nikkor 105mm
Kodachrome 64 • f4 @ 1/250

"This right-breaking Shark Island wave came right through the pack on a very grey day, and nobody wanted it except the photographer. Having no board, I did the next best thing and captured the beast's image. A favorite shot of mine!"

CXV
Maalaea, Maui, Hawaii
KIRK AEDER
Canon A-1 • 35-70mm zoom
Fujichrome 100 • f8 @ 1/250

"Obviously, the comparison here to a set of ocean waves is inevitable. That's the only reason I took the picture!"

CXVI
West Palm Beach, Florida
TONY ARRUZA
Nikon F2 • (Unrecorded telephoto)
Kodachrome 64 • (Unrecorded)

"A beautiful Florida winter sunset created the purple and blue colors in the sky, which reflected onto the ripples of the brown water."

CXVII
Sunset Beach, Oahu, Hawaii
WARREN BOLSTER
Canon F-1 (High Speed) • Century 1000mm
Kodachrome 64 • f6.8 @ 1/500

"Corduroy to the horizon… power-sweeping Sunset. Shot from part way up Comsat bill. Shot an entire roll on this set as I'd never seen anything like it! Probably a 12-to-15-foot day at Sunset. A rare moment in time…where the power lines give a tranquilizing effect. One of this series hangs on my wall at home; only three photos do."

CXVIII
Moss Landing, California
WOODY WOODWORTH
Nikon FE • Nikkor 20mm
Kodachrome 64 • f8 @ 1/1000

"Moss Landing is in the center of Monterey Bay. The sun was so glaring on this angle when flying over, I had to really underexpose to hold on the line patterns."

CXIX
San Simeon, California
GUY MOTIL
Contax RTS II • Zeiss 300mm
Kodachrome 64 • f8 @ 1/500

"Offshore wind texture mixed with sunlight."

CXX
Waimea Bay, Oahu, Hawaii
MOE LERNER
Nikon 2020 • Nikkor 50mm
Fujichrome 100 • f8 @ 1/500

"Who would think that I, Moe, would try to take a watershot of the Waimea shorebreak? But it was only three feet."

CXXI
North Shore Oahu, Hawaii
DENJIRO SATO
Nikon FE2 • Nikkor 85mm
Kodachrome 64 • f8 @ 1/500

"Banzai Pipeline. What can you learn from the ocean? Almost everything. It is like a religion to me. I need it. It is always stronger than me. It has taught me that life is the most important thing. It makes my life simple, but sometimes confusing, too."

CXXII
Zuma Beach, California
STEVE SAKAMOTO
Canon F-1 • Canon 800mm
Kodachrome 64 • f8 @ 1/500

"Steve Sakamoto lived in Long Beach, California, and his photo range was south to Huntington Beach and north to the Ventura/Oxnard area. Early morning sun-dazzled images are one of his trademarks."
—Steve Pezman

As we should know from the study of undulatory vibrations in the world of physical phenomena, every wave comprises in itself a complete circle, that is the matter of the wave moves in a completed curve in the same place and for as long as the force acts which creates the wave. We should know also that every wave consists of smaller waves and is in its turn a component part of a bigger wave. If we take, simply for the sake of argument, *days* as the smaller waves which form the bigger waves of years, then the waves of *years* will form one great wave of *life*. And so long as this wave of life rolls on, the waves of days and the waves of years must rotate at their appointed places, repeating and repeating themselves. Thus the line of the fourth dimension,

the line of life or *time*, consists of wheels of ever-repeating *days*, of small circles of the fifth dimension, just as a ray of light consists of quanta of light, each rotating in its place so long as the primary shock which sends forth the particular ray persists. But in itself a *ray* may be a curve, a component part of some other bigger wave. The same applies to the line of life.

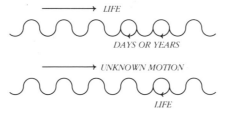

If we take it as one great wave consisting of the waves of days and years, we shall have to admit that the line of life moves in a curve and makes a complete revolution, coming back to the point of its departure. And if a day or a year is a wave in the undulatory movement of our life, then our whole life is a wave in some other undulatory movement of which we know nothing. As I have already pointed out, in our ordinary conception life appears as a straight line drawn between the moments of birth and death. But if we imagine that life is a wave, we shall get this figure:

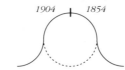

The point of death coincides with the point of birth.

— *P.D. Ouspensky*
A New Model of the Universe

PRODUCTION NOTES

Project Directors: Patrick O'Dowd and Steve Pezman
Editorial Director: Drew Kampion
Art Director: Jeff Girard
Photography Editor: Art Brewer
Illustrator: Phil Roberts

Handlettering: Paul Kulhanek
Copy Proofing: Jody Kirk
Production Assistance: Mark Sansom
Publishers' Assistants: Denise Bashem and Chris Lyons
Photo Services: Tom Servais and Bill Dewey
Photography Editor's Assistant: Rob Gilley

Typography: Set in Adobe ITC Garamond on Macintosh electronic design equipment.
Output on Linotronic 300 Imagesetter at Central Graphics, San Diego, California.
Chapter titles were set in a specially-drawn version of Garamond bold condensed.
Color Separations: Four-color separations made at AGEP in Marseille, France
on an 399 TE Hell scanner at 175 lines per inch with Laser Scanner film S711p from AGFA-GEVAERT.
Proofed on Chromalyn equipment from Du Pont de Nemours.
Printing: four-color sheetfed offset printing and binding done at PROOST at Turnhout, Belgium
on 170 gram Satimat supplied by ARJOMARI DIFFUSION.

Jacket design by Jeff Girard
Endpaper photo of water ripples by Sato